Why is understanding the design and manufacturing requirements of footwear tooling so important?

Simply put, tooling is essential to building a modern footwear brand. Tooling is expensive and time consuming to develop. Get it right and your brand will grow, get it wrong and your brand is heading for tough times.

SHOEMAKERS ACADEMY

Online Shoemaking Courses
for all levels of Shoemakers,
Shoe Designers, Developers, and
Brand Builders.

Expand your skills and knowledge in modern shoe design, manufacturing, sourcing, and footwear brand development.

Jumpstart your career. Launch your shoe brand.

Available worldwide, on-demand,
Online, and on your schedule.
Start Today!

Textbooks Available now:

How Shoes are Made
Footwear Pattern Making and Last Design
Shoe Material Design Guide
How to Start Your Own Shoe Company
How to Spot Fake Sneakers
Footwear Tooling Design Guide
Como Os Calçados São Fabricados
Cómo se hacen los zapatos
Cómo empezar tu propia empresa de calzado
Guía para el diseño de materiales de calzado
Patronaje de calzado y diseño de hormas
Giày được sản xuất như thế nào
鞋子是怎样制成的

Online Shoemaking Courses

Shoemaking for Designers and Brand Builders
How to Select Shoe Materials
Shoe Business Start-Up Course
Footwear Tooling Design and Development
Sneaker Authentication Basics
Creating Footwear Specifications
Footwear Import Duty Rules and Regulations
Sustainability Strategies for Footwear
Footwear Cost Calculation
How to Communicate with Shoe Factories
PROcreate & Sketching for Shoe Designers
Build it with Gavica: Air Jordan 1 High
Make It Custom with Rob: Painting for Beginners
Footwear Cost Engineering
Footwear Inspection and Quality Control
Building a Modern Shoe Factory
Footwear Sustainability Strategies
Footwear Fitting & Comfort
Adobe Photoshop For Shoemakers
Footwear Import Duty

Footwear Tooling Design Guide

Written and edited by
Wade and Andrea Motawi
ISBN-13: 978-1-7358833-6-6

TheShoeDog@ShoemakersAcademy.com

Enjoy!
Special thanks to:
Andrea, Alex, Erik, Karim, Halla, Mom, Kamal, Karim, Joanne, Dave, Alfredo, Jason, David, Lizzie, Johnson, Steve, Lenny, Bernie, Jenny, Simon, Ben, David, Leo, Jack, Gary, Jean Marc, Suresh, Abdón, Paolo, Patricio, Andres, Lou, Fernanda, Rafaela.

Thanks to all my working friends in the USA, China, Ecuador, Hong Kong, Korea, Taiwan, Vietnam, and Europe.

Footwear Tooling Design Guide

The Complete Guide for the Design, Development, Costing and Manufacturing of Footwear Tooling.

WADE MOTAWI

FOOTWEAR TOOLING DESIGN GUIDE

Chapter 1

Footwear Tooling Design, Development, & Costing.

Why it's important to make your own footwear tooling.

Your first great tooling decision!
Our goal for this book and online course at ShoemakersAcademy.com is to prepare you to make smart tooling decisions for your footwear brand. You are reading this book so you are well on your way.

With this guide you will learn:
The tooling required to make modern shoes.
How to design tooling.
Tooling development processes.
Equipment manufacturing costs.
Tooling and supplier relationship management.

We will review the specific equipment you need to manufacture YOUR shoe design. We will try not to talk about the equipment that is standard to the factory, but focus on what the factory needs to make YOUR unique shoe designs.

Once you have a clear understating of the equipment types, design parameters, development processes, and tooling management techniques, you will be able to avoid costly mistakes.

Equipment basics
To manufacture footwear you're going to need tooling. Sometimes we call it equipment. What are you going to need?

1. Shoe lasts
2. Material cutting dies
3. Outsole and midsole tooling

Depending on the design, you may need injection molds, high frequency welding tools, and more.

We will walk you through these requirements step-by-step and dive deep into the information so you can look at your design and figure out what tooling is required and how much it will cost.

Let's go!

Don't Be Intimidated By Tooling.

By the end of this book you will be able to handle important tooling decisions tooling with confidence.

You will:
Identify tooling types,
Understand your tooling requirements,
Build equipment budgets, and
Know how to manage your tooling.

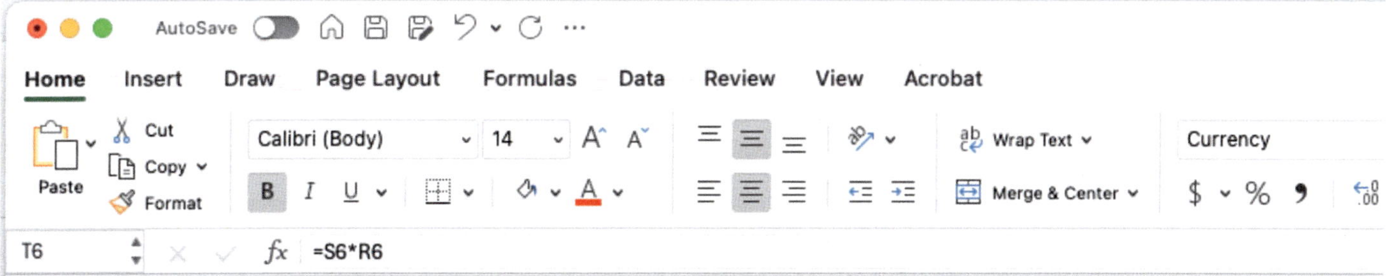

Shoe Development Project Pull over Phase 1

Shoe Size		7	7.5	8	8.5	9	9.5	10	10.5	11	11.5	12	12.5	13	14			
Sample Room Fee Overhead																1	$500.00	$500.00
Model Last Development	NEW LAST					1										1	$555.00	$555.00
Lasts needed for testing						4										4	$35.00	$140.00
Upper Pattern Work						1										2	$400.00	$800.00
2D CAD WORK						1										1	$175.00	$175.00
3D CAD work						1										1	$400.00	$400.00
PU CASTINGS OUTSOLE / MIDSOLE						1										1	$400.00	$400.00
Finished Fit Test samples						4										4	$72.00	$288.00

$3,258.00

Shoe Development Project First Size Mold Making Phase 2

Sample mold cost		7	7.5	8	8.5	9	9.5	10	10.5	11	11.5	12	12.5	13	14			
Outsole Compression Rubber						1										1	$1,425.00	$1,425.00
Midsole Compression EVA						1										1	$1,500.00	$1,500.00
Midsole Injection						1										1	$5,000.00	$5,000.00
Upper molds						1										1	$300.00	$300.00
Screen Prints						1										1	$100.00	$100.00
EVA Footbed						1										1	$425.00	$425.00
Sole Pressing Fixture						1										1	$75.00	$75.00
Cutting dies						1										1	$150.00	$150.00
Working prototypes	Number TBD					1										8	$125.00	$1,000.00

$7,050.00

Shoe Development Project Production Equipment Mold Making Phase 3

Size Run US Mens	Type	7	7.5	8	8.5	9	9.5	10	10.5	11	11.5	12	12.5	13	14			
Outsole Compression Rubber		1		1				1		1		1		1	1	7	$1,425.00	$9,975.00
Midsole Compression EVA		1		1				1		1		1		1	1	7	$1,500.00	$10,500.00
Midsole Injection		1		1				1		1		1		1	1	7	$5,000.00	$35,000.00
Upper molds		1		1				1		1		1		1	1	7	$300.00	$2,100.00
Screen Prints		1		1				1		1		1		1	1	7	$100.00	$700.00
EVA Footbed		1								1				1	1	4	$425.00	$1,700.00
Sole Pressing Fixture		1		1				1		1		1		1	1	7	$75.00	$525.00
Cutting dies		1	1	1	1		1	1	1	1	1	1		1	1	12	$150.00	$1,800.00
Production lasts	Plastic	12	12	12	12	12	12	24	24	24	12	12	12	12	12	204	$13.00	$2,652.00
Credit for 2D and 3D Cad work																0		$575.00
Production Cutting Dies	Metal	1	1	1	1	1	1	1	1	1	1	1	1	1	1	14	$150.00	$2,100.00

$66,477.00

Shoe Production Cost Phase 4

Size Run US Mens	Type	7	7.5	8	8.5	9	9.5	10	10.5	11	11.5	12	12.5	13	14	Pairs	FOB Cost	Total
																1200	$ 28.00	$ 33,600.00
																Shipping	1.5	4500
																Service	$ 2.24	$ 2,688.00
																Duty	$ 4.76	$ 5,712.00
																Landed	$ 36.50	
																Total	COGS	$ 43,800.00
																	Tooling + Dev	$ 76,785.00

Building Tooling Is How You Build A Modern Brand!

Iconic shoe designs were made possible by manufacturing advancements.

That means they needed new tooling.

Tooling is how brands deliver technical innovations

Tooling is how brands deliver new material innovations

Tooling is how brands deliver new performance products

Tooling is how brands deliver wild styles

Tooling can be your entire brand

Tooling is how brands deliver value and efficiency

Creating Footwear Tooling is a Team Effort

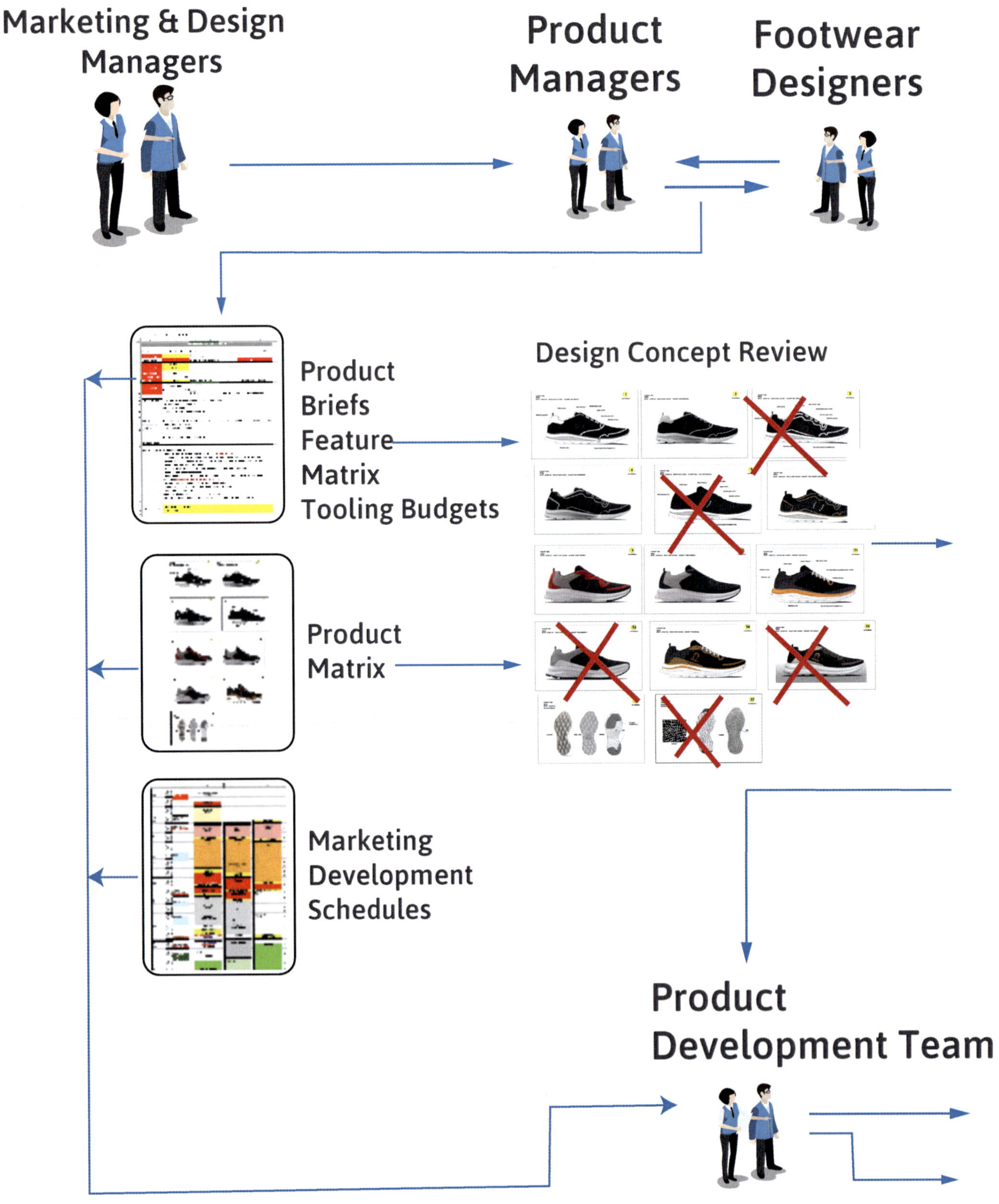

Creating shoes and tooling is a team effort. Everyone involved in the development and brand management teams needs to be engaged. Let me explain why.

This chart illustrates how product managers and design managers collaborate with footwear designers.

Together, they develop product briefs, marketing briefs, and schedules. They create a line plan, select items, and convince managers that a project is viable. Part of this process includes deciding what technology to incorporate. All these decisions funnel down into product development, where budget constraints and technological requirements are turned into technical specifications before reaching the shoe factory.

Once everything reaches the factory, another wave of people get involved because, as you'll find out, shoe factories make shoes—they don't make tooling. Tooling is outsourced, so many different parties get involved, but that's okay.

We're going to walk through how that works.

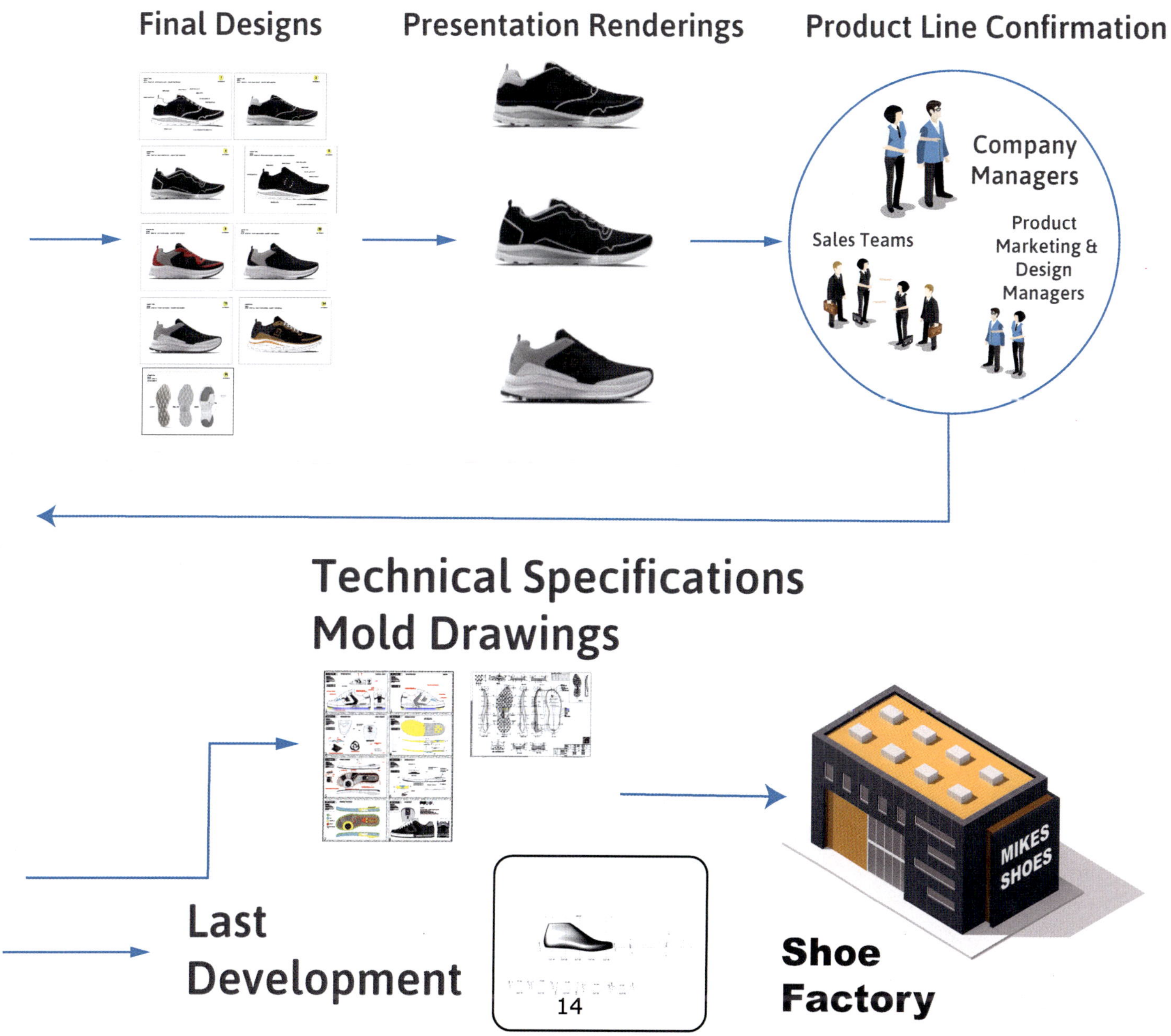

Footwear Designers, I'm talking to you!

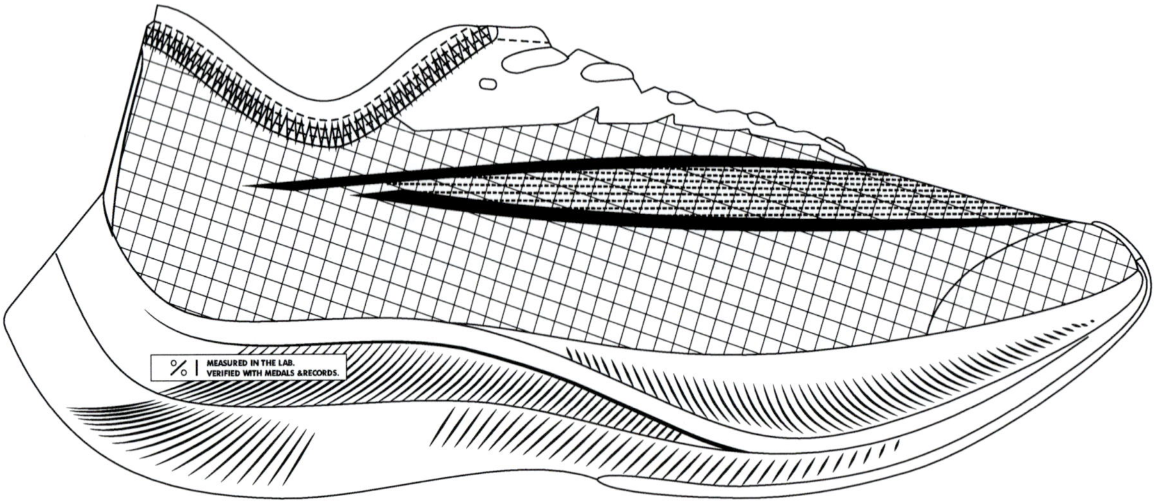

Footwear Developers and Product Managers,
You can grab a coffee for a moment.

Footwear designers, this part is for you.
As a designer myself, I say this with respect: it's crucial for you to understand the limitations of the equipment.

I want you to unleash your creativity, but you need to know when you're pushing developers to do something truly new and when you're making an impossible request. It's a matter of time and feasibility. If your product manager tells you that you have two years to develop a new technology, that's great. If they say you have two months, you'll need to approach things differently.

Designers, it's your job to challenge developers and product managers to create innovative, interesting products. However, understanding how the equipment works will help ensure your designs can actually be made. If you design something that's impossible to produce, your original intent will be lost in the translation.

Yes, shoe designers, this is really important. You need to know this stuff so you can create amazing designs...that can be made.

Of course, it's a designers job to challenge the developers and tool makers, but if you understand how this equipment works, and how much it costs, then you can make better designs.

Footwear Developers, I'm talking to you!

Developers, now it's your turn.
I moved from design to development and eventually became a development manager. As a developer, it's your role is to figure things out.

Designers and product managers will bring you ambitious concepts and you need to collaborate with factories and tooling vendors to make them a reality. They're counting on you to connect the dots.

As a developer, you must dive into details like:

How much time will it take?
How much will it cost?
Who can make this mold?

You may receive an amazing design and a beautiful concept, but if it's not feasible, it's your responsibility to communicate that. You need to assess whether it can be made, and if so, at what cost.

Designers and product managers are counting on you to deliver the design intent on time and on budget.

Developers, it's your job to figure it out!

Product Managers, I'm talking to you!

Product managers, you're steering the ship.
You provide instructions to designers and developers, and it's your plan that guides the entire process.

Tooling is expensive and time-consuming to develop, so you need to determine whether a project is worth the investment.

Do you have the budget and time for it?
Can you use existing tooling?
Do you need new molds?

It's also your job to coordinate efforts between designers, developers, and the factory team.
Ultimately, product managers are responsible for the plan. Depending on your company structure, the marking team may need to have a say too.It may be the marketing team that is helping set the product direction.

Business category managers, you're involved too. If you're running a small company, the responsibility falls on you to figure everything out.

Product managers, you are responsible for the plan.

How Do We Get From Ideas To Tooling?

Tooling Development Process

1. Plan your tooling budget.
2. Create your shoe last.
3. Create the upper sample pattern pull-over.
4. Scan completed pull-over.
5. Use scan data to create cross sections of the 2D blueprint.
6. Review 2D drawings, confirm the design and part engineering.
7. Create 3D CAD file, reconfirm the design and engineering.
8. Create a solid model (woodcut) by 3D print or CNC cut.
9. Create temporary cast resin tooling (optional).
10. Confirm 3D file and solid model changes.
11. Send to metal casting, CNC factory, or 3D print supplier.
12. Raw metals casting is scanned.
13. Casting is finished by CNC then assembled.
14. Test pressing - test parts are stock fit to confirm dimensions.
15. Mold texture is added.
16. Teflon coating is applied to protect the tooling.
17. Mold is complete and ready for final testing.

Chapter 2

Tooling Cost and Productivity Calculations

Tooling may be the largest capital asset of your young brand. Great tooling can last a brand 40 years. Bad tooling can fail after one season.

You will need to consider your sales forecast versus the tooling cost and mold productivity.

Tooling, Pricing, and Productivity in Footwear Production

Tooling plays a crucial role in footwear manufacturing by directly impacting costs, efficiency, and overall production capabilities. Understanding the unit cost of tooling, the number of sizes required, and productivity is essential for managing a footwear production process effectively.

Tooling Costs and Investment Considerations

Footwear tooling is a significant investment. A basic sneaker can require an outsole mold costing approximately $1,500 per size, while a complex design with injection molds and advanced components may exceed $10,000 per size. Consequently, a standard tooling setup may cost as low as $25,000, whereas a highly specialized shoe could demand an investment of $150,000 or more.

For major brands such as Nike, tooling costs amount to millions of dollars, a manageable expense given their scale. However, for smaller brands, tooling represents a major capital investment that must be carefully tracked and protected against damage or loss. In some cases, lost or damaged tooling has resulted in production setbacks that small companies could not afford to remedy.

Longevity of Tooling

High-quality tooling can last for years. Some brands continue to use molds from the 1980s. However, tooling may become obsolete due to design changes, market shifts, or technical issues sometimes rendering an entire size run unviable before production even begins.

Sales Forecasting and Tooling Investment

A brand's sales forecast is a key factor in tooling decisions. Small brands with niche products must balance cost constraints with production needs. Some large brands, despite anticipating low sales volumes for certain models, still invest heavily in tooling.

The choice between higher-cost, high-productivity tooling and lower-cost, less efficient tooling depends on production scale and timelines. Some brands may opt for more expensive molds to increase production speed, while others prioritize cost savings over efficiency.

Building Equipment Budgets

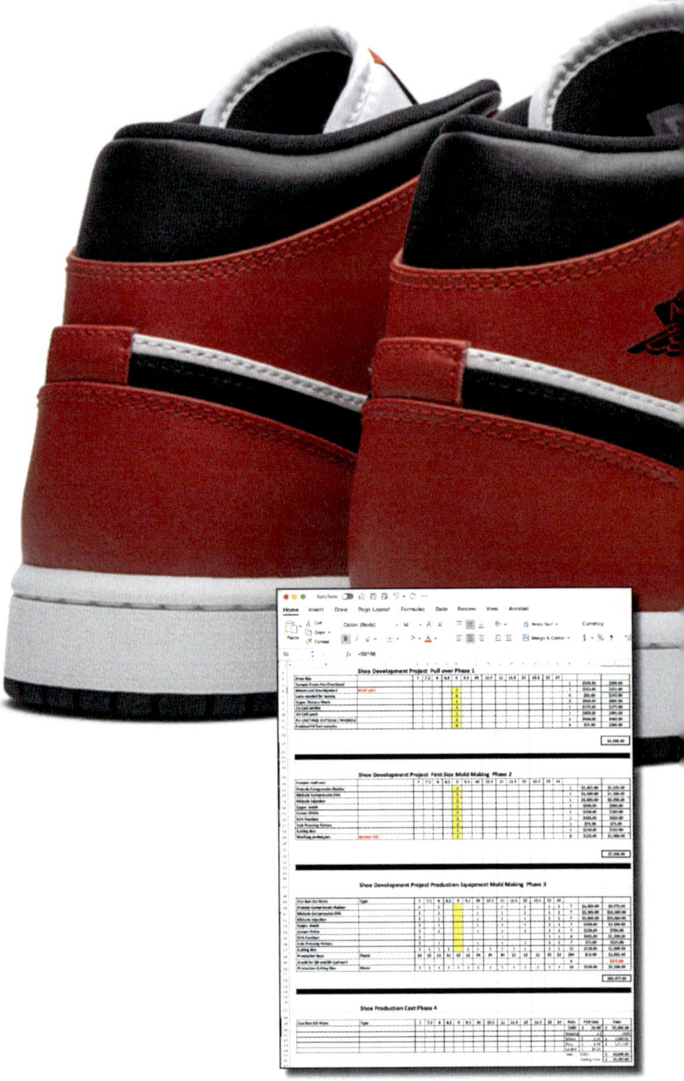

For most managers, making tooling budgets isn't much fun. How do you estimate the costs for the tooling of items that haven't been designed yet? This seems like an impossible task.

It's not. We will review the most common and uncommon mold types. By the time you are finished you will be able to handle it.

Don't forget, it's okay to ask for help. Your factories and mold makers are standing by to help you launch new projects.

A Simple Budget

Nike Air Jordan 1
Tooling Cost

18 Sizes = $54,750

NOT a Simple Budget

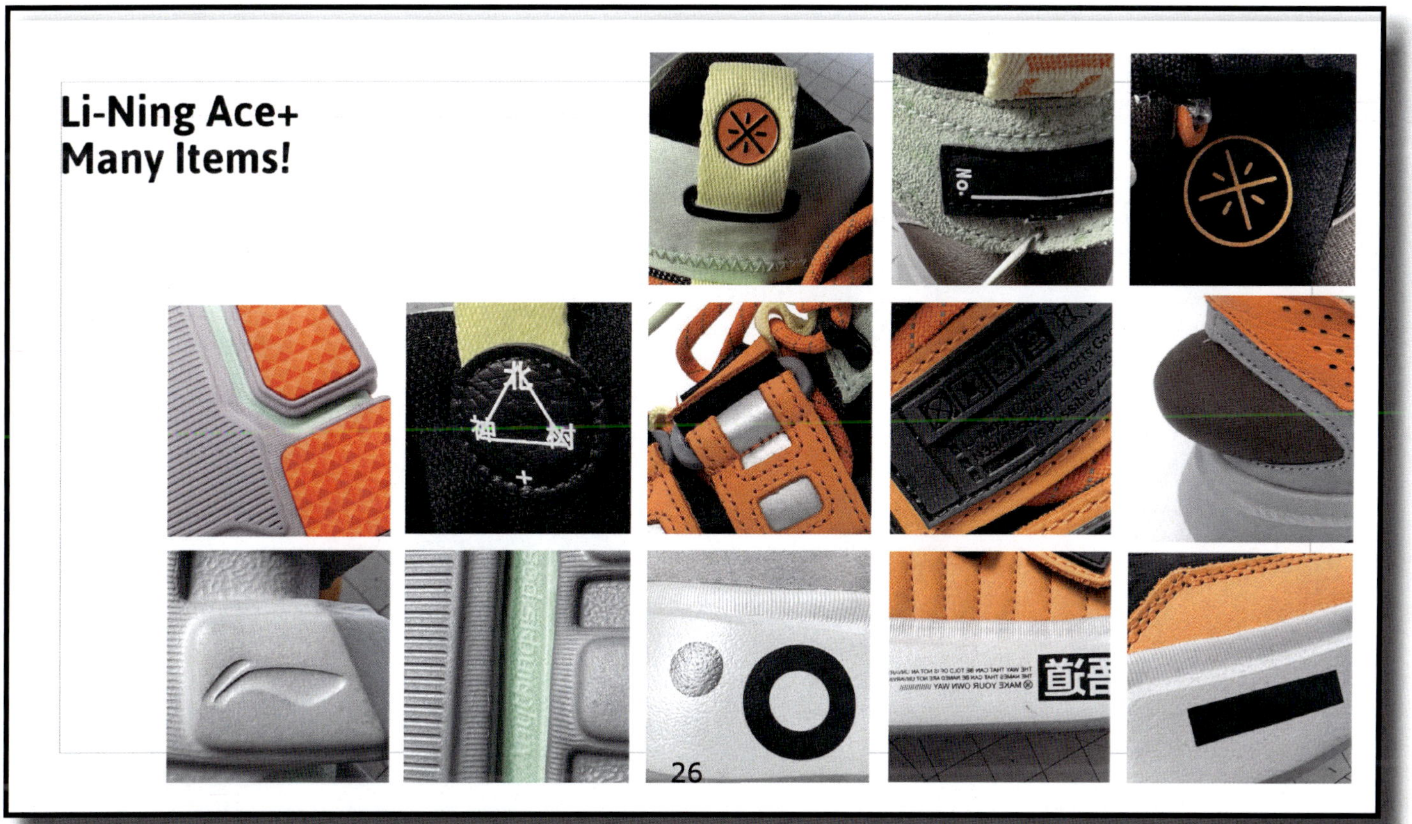

boost
boost

Tooling Costs

3D injection mold
5 sizes x $5500 = $27,500

2D - 2 part injection mold
5 sizes x $4000 = $20,000

Footbed mold
12 sizes x $750 = $9,000

Midsole mold
12 sizes x $3,500 = $42,000

Midsole shank 3D mold
6 sizes x $4,000 = $24,000

Rubber outsole mold
12 sizes x $1,500 = $18,000

Total = $140,500*

*Does not include duplicate size molds

Tooling Size Requirements

How many of each tool do you need? One for each size? Maybe not....

The 3D heel injection mold is an expensive 3D mold. Due to the last heel width standardization and mold productivity, it is possible to group grade and share this part. Maybe 4 sizes is okay? For example, 7, 7.5, 8, and 8.5.

This logo eyestay side panel is a dual injection part that could be molded flat or in 3D. This is a relatively expensive injection mold so the brand would like to share or group grade across a few sizes. Because this part floats on the side of the shoe and does not interfere with the pattern there is more opportunity to share.
Maybe 4 sizes is okay? For example, 7, 7.5, 8, and 8.5.

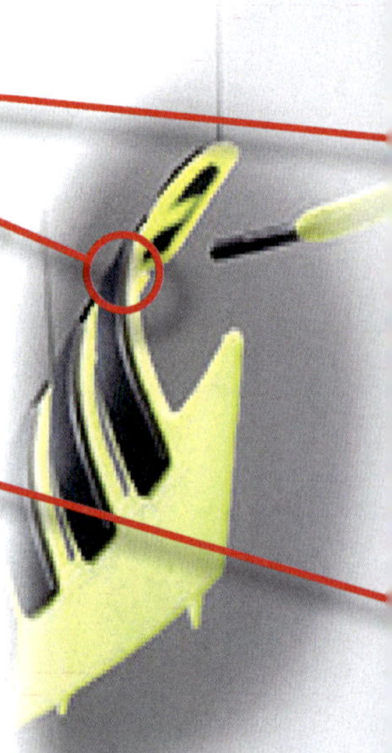

Footbed tooling is generally inexpensive, cold pressed EVA parts with die cut edges. Again, due to last heel width standardization one mold can usually support 4 to 6 sizes. This design, however, is not die cut so it must be molded per size. This is uncommon and will increase the footbed tooling and unit cost.

Midsole molds can be shared across sizes but only if they are designed for this purpose and the shoe design allows it. For example, this design could use a size 7.25 and lasts sizes 7 and 7.5. However, midsole tooling is not as productive as rubber pressing or injection molds, so a brand like Adidas would need more molds to support their production schedule. Thus, there is no cost savings.

The injected shank component can be shared across a few sizes, again depending on the design details. Injection molds have a high rate of production so there are no capacity issues. Also, as the shank is integrated with the midsole, the midsole interface with the shank can be group graded even if the midsole size itself is modeled 1 to 1 with the lasts.

The rubber outsole mold is usually designed to follow midsole sizes 1 to 1. However, if the rubber parts are inserts or pods the midsole can be designed to accept group graded rubber parts.

boost
boost

Footwear Tooling Mold Amortization

Tooling can be very expensive.
Use amortization to pay it off?*

What is the "cost" of money?
Interest payments?
Relationship with the factory?
Import duty is a HUGE factor.

***Maybe**

The factory and customer agree to pay for the tooling piece-by-piece by adding the cost to each shoe.

For example:

50,000 pairs x $2.00 per pair = $100,000

1,000,000 pairs x .10 per pair = $100,000

Tooling Amortization

Mold amortization allows brands to finance tooling by incorporating its cost into the per unit price of the shoe. A factory may agree to recover mold costs over time by adding a small fee, such as $1 per pair, until the tooling cost is fully paid. **However, amortization comes with drawbacks**:

Interest Rates:

Factories are not banks, and interest rates on amortization agreements may be high. In the old days, factories did not charge interest....but today it's a different story.

Import Duties:

If tooling costs are built into the per unit price, import duties which apply to every pair will increase the overall expenses. Shoe brands are required to pay duty on "assists." An assist is the tooling used to produce the shoe. However, the "assist" duty is generally only paid on the first delivery of the shoe. You will need to follow up with your import and duty specialists.

Factory Constraints:

Some factories may not have the cash flow or financing options to offer amortization, especially to new brands.

Due to these factors, many brands prefer to pay for tooling upfront rather than amortizing costs over production runs.

It's very rare for new brands to get amortizing deals.

Footwear Mold Productivity

Different tooling and production methods significantly impact equipment costs, mold productivity, and product performance.

Brands must consider the projected life of the design and tooling to fully understand the return on investment.

Production Efficiency and Cycle Times

In manufacturing, production cycle times vary based on mold type and machine efficiency. For example:

Direct-injection machines can produce a component in 15 to 20 seconds per cycle.

PU foam molding can produce parts every three seconds.

Rubber pressing molds can produce parts every five minutes.

EVA compression molds can produce one part every 10 minutes.

Understanding these cycle times helps brands balance tooling investments with production efficiency.

Brands must carefully evaluate tooling investments based on their budget, sales forecasts, and production requirements.

Productivity varies across tooling methods with injection molding offering faster production but at a higher upfront cost.

Manufacturing scale determines mold duplication needs ensuring timely delivery and production efficiency.

By strategically managing tooling investments, brands can optimize production efficiency, maintain cost control, and achieve long-term profitability in the competitive footwear industry.

PU Direct Attach

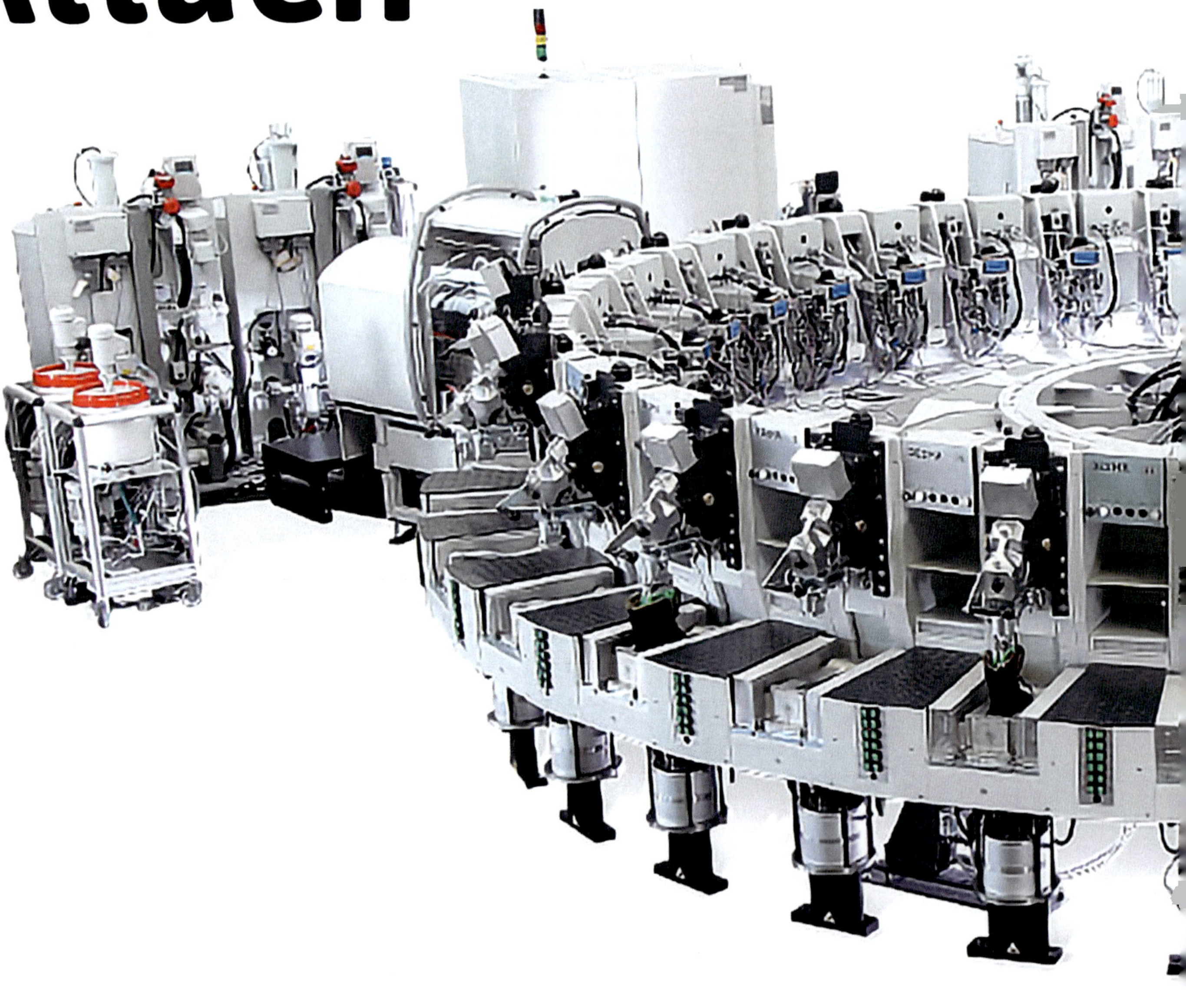

Expensive tooling
High productivity
Very expensive machine
Great product

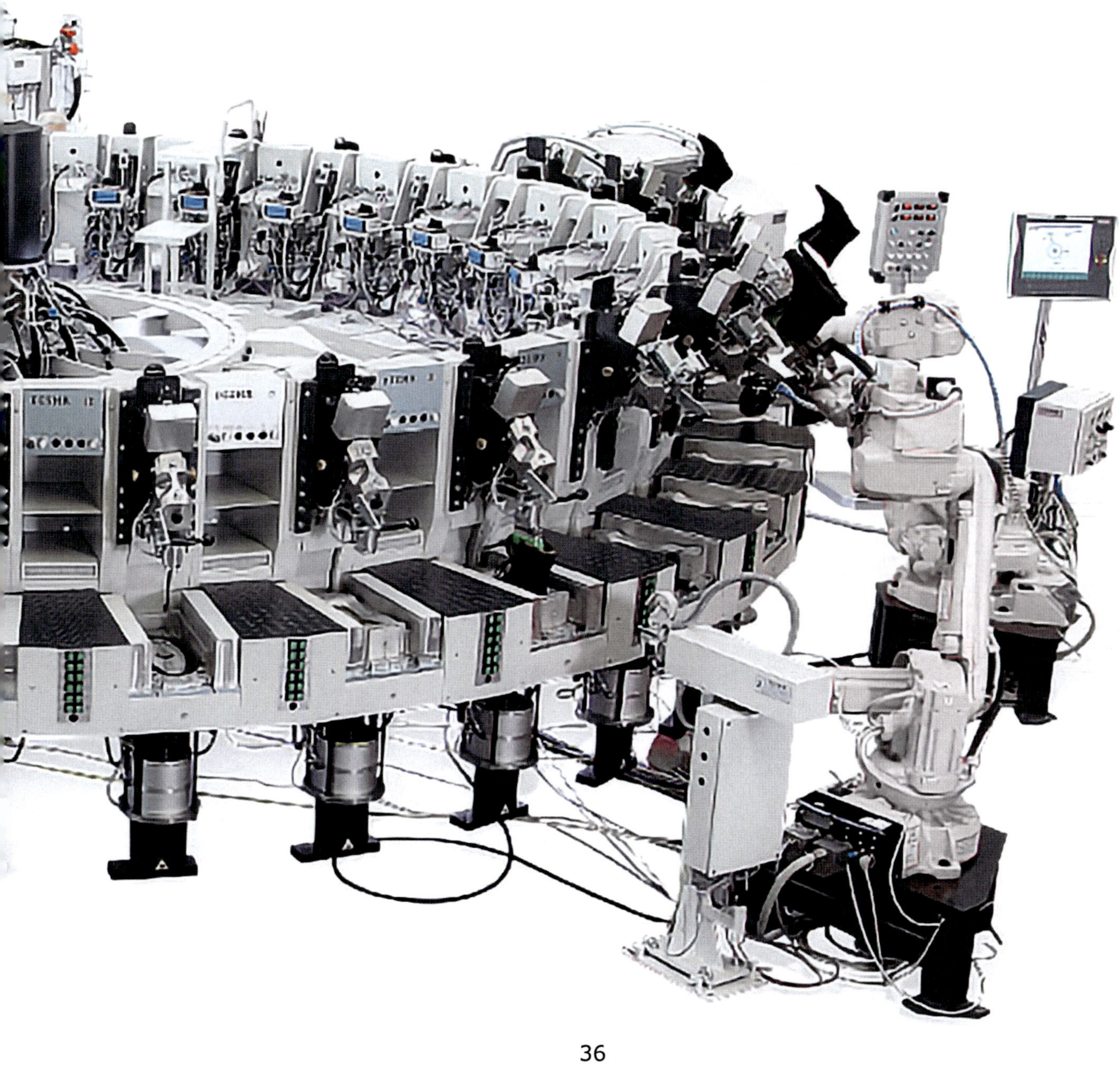

PVC Direct Attach

Expensive tooling
High productivity
Inexpensive machine
Basic product

Chapter 3

What About Public Molds Or Open Molds?

"Hey Shoe Dog, what about public or open molds?"

Open molds, refers to existing equipment that has already been created and is available for use. Because they are complete there's no waiting period to access them. That may sound great, but there are some important issues to consider.

Key Issues with Open Molds

What last was used to create the tooling?
If you're building your own brand, you likely have a distinctive shape. The last used in the tooling determines that shape. If it's not your specific last, you won't be able to use the tooling effectively.

What is the size run?
Does the mold cover the full range of sizes you need?

What is the size scale?
Many open molds were originally designed for the Chinese domestic market. If you're selling in Europe or the U.S., the size increments may not align properly. For example, European sizing doesn't include half sizes which could create holes in your size run.

Where is the tooling located?
If you find a mold on Alibaba, remember—China is a big country. If you're producing in Vietnam but the mold is in China, it's not a viable option.

The tooling won't have your logo.
Some molds may have removable plugs for branding but they are not designed specifically for your company. The design won't be unique to you.

You don't control the equipment.
If the supplier decides they no longer want to sell it to you, or if the relationship between your assembly factory and the tooling factory deteriorates, you could lose access to the mold entirely.

Who is making the parts?
Is the mold sitting at a rubber presser or with the mold maker? Do you have control over who will be the presser or injector? This is not likely.

What about quality control for the finished components?
Who controls the material specifications? Do you want a local Chinese market spec, or do you need something more durable or slip-resistant?

Can you use the parts at a different factory?
This is often a real challenge. Suppose you're working with a factory that you love but they can't make waterproof shoes and you want to use the same outsole unit at a different facility. If the tooling isn't transferable, you're stuck.

What about mold maintenance?
I've seen products arrive with speckles inside, only to realize the mold was rusty because no one maintained it properly.

Why having your own equipment matters

The last is your design.
The last is made to your specifications ensuring the shape and fit is exactly what you want.

You control the size run and scale.
The size run is made for your market and the size scale will meet your customer expectations.

The tooling location
This is your equipment. You can move it wherever you want whenever you want. You can also ship parts to whomever you want.

Your branding
This is your design with your logo and you don't have to share it with anyone.

Production flexibility
You can choose which factory makes your parts and where the parts are shipped.

Part suppliers
You can select who makes the components and this insures you can maintain a direct relationship with the supplier. No middle men.

Quality control
The molds are in your control and you can inspect the equipment and parts any time you like.

Material specifications
Again, if you own the molds and maintain the relationship with the supplier, they must answer to you.

Mold maintenance
You can proactively address wear and tear.

Better safe than sorry.
When you own the mold, you're not at the mercy of a supplier who may restrict access or refuse to sell to you.

Tooling Is A Huge Investment. Let’s Do It Carefully.

If your goal is to be a REAL brand...

Don't use "open" molds*.

*unless it's Vibram

The Exception: Vibram

One possible exception is Vibram, a reputable supplier with molds in both China and Italy.

Vibram offers open molds but some of the same concerns apply. They also design custom molds for clients which is a great option if you can afford it.

Even with Vibram, there are still questions:

What last was used to create the tooling?

What is the size run?
Usually UK or European sizing.

Can it work for U.S. sizes?
Sometimes I've successfully adapted European Vibram molds for U.S. sizing.

Where is the tooling located?
Probably in China or Europe but you do not own it.

Does it have your logo?
No—it has Vibram's.

Do you control the equipment?
No.

Who makes the parts?
Vibram. They make high-quality components.

What about QC?
Vibram ensures good quality.

Who controls the material? Specifications?
Negotiated with Vibram. They maintain high standards.

Conclusion
While Vibram offers a solid option for some, if you are building a real brand with a unique design and logo, avoid open molds.

In fact, when clients come to us, we rarely take on open mold projects because of these complications.

Chapter 4

The Shoe Last Is Equipment

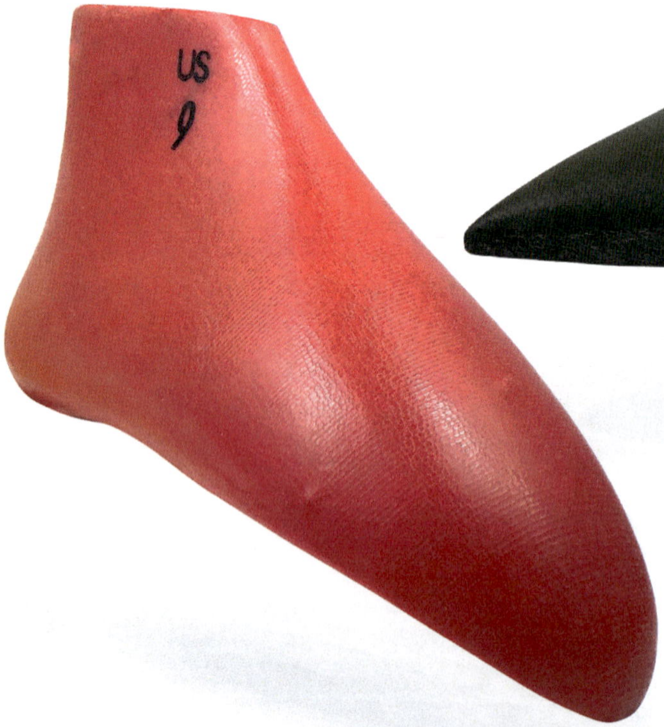

The shoe last
is a critical part of your
equipment.

The Importance of the Shoe Last in Footwear Production

The shoe last is a critical component in footwear manufacturing which is often overlooked as a piece of equipment. However, it is essential to the design, fit, and function of a shoe, making it an indispensable investment.

The shoe last is unique to each brand, style, and model, serving as the foundation for shoe construction.

The Role of the Shoe Last

A shoe last is crucial to shaping and defining the characteristics of a shoe, including its fit, style, and functionality. It is a fundamental tool that dictates how the shoe conforms to the foot.

While some factories may offer generic or open lasts, it is advisable to develop a custom last to ensure product differentiation and superior fit.

The last does not perfectly replicate the human foot; rather, it is stylized and adjusted for performance. Whether the goal is aesthetics, traditional design, or athletic functionality, each shoe last must be tailored to meet specific performance needs.

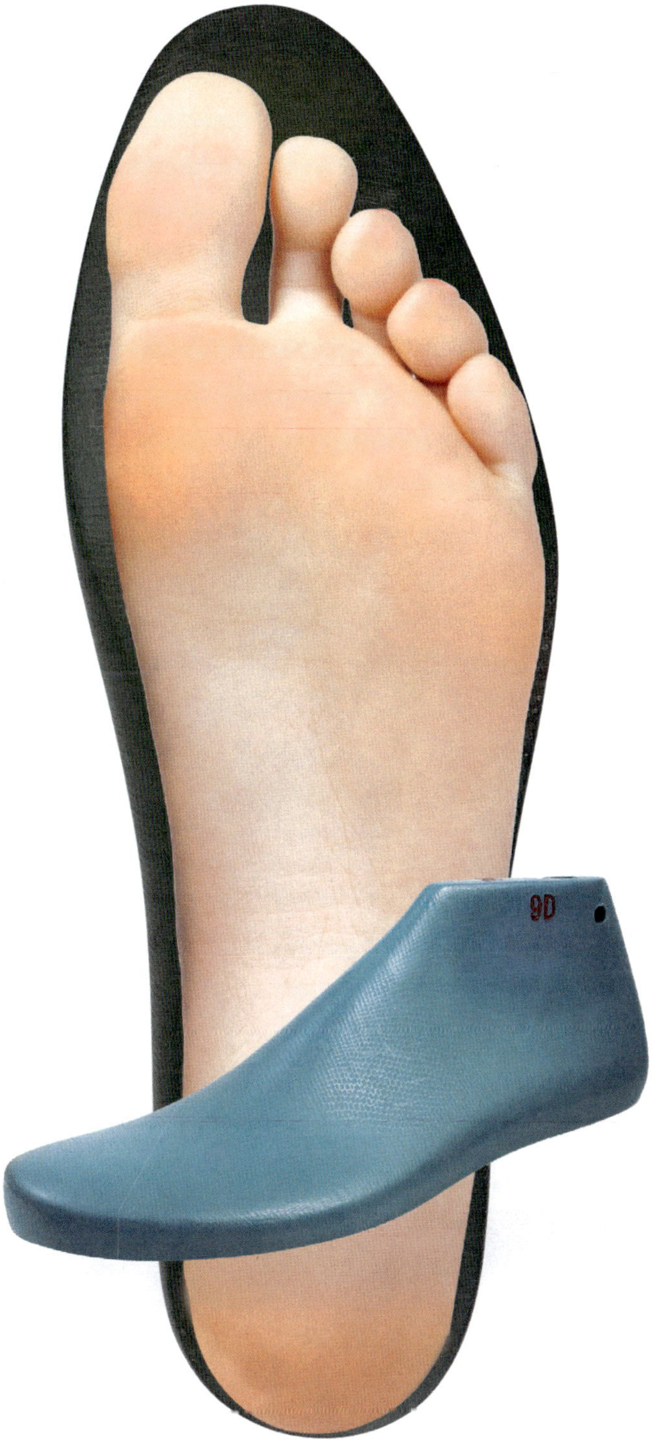

Development of a Shoe Last

Creating a shoe last involves both high-tech processes and skilled craftsmanship. Typically, a blueprint is developed based on a foot scan, followed by manual refinements to perfect the contour.

Technicians sculpt the last to achieve the desired shape before digitizing it into a 3D computer model for further precision. This ensures that the last accommodates the human foot while maintaining the intended design elements.

A 3D model provides valuable insight into the geometry of the last, but physical measurements, such as length, width, ball girth, and instep height, must still be verified. These dimensions directly influence the shoe's comfort and performance.

DONG GUAN BANGYIDA SHOES MATERIAL FACTORY

CUSTOMER: 88Manufactnring LLC **SIZE :** 35#-45#

STYLE : PISA SP LT-1

GRADING :L1-L2:6.66mm E1-E2:4.75mm

C1,C2=T1,T2= **DATE:** **2025-02-11**

MEASUREMENT/SIZE	INCR		35#	36#	37#	38#	39#	40#	41#	42#	43#	44#	45#
BOTTOM LENGTH (L1-L2)	6.66		230.72	237.38	244.04	250.70	257.36	264.02	270.68	277.34	284	290.66	297.32
BOTTOMWIDTH (A1-A2)	1.58		79.36	80.94	82.52	84.10	85.68	87.26	88.84	90.42	92	93.58	95.16
Tread Girth (E1-E2)	4.75		216.00	220.75	225.50	230.25	235.00	239.75	244.50	249.25	254	258.75	263.50
WAIST Girth (F1-F2)	4.75		218.00	222.75	227.50	232.25	237.00	241.75	246.50	251.25	256	260.75	265.50
INSTEP Girth (G1-G2)	4.75		218.00	222.75	227.50	232.25	237.00	241.75	246.50	251.25	256	260.75	265.50

COMMENTS

FOOTJOY-43#

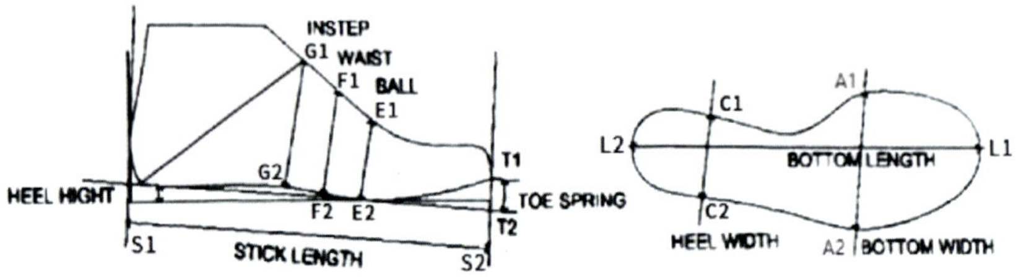

Sheet1 Sheet2 Sheet3

Last making is an art and science

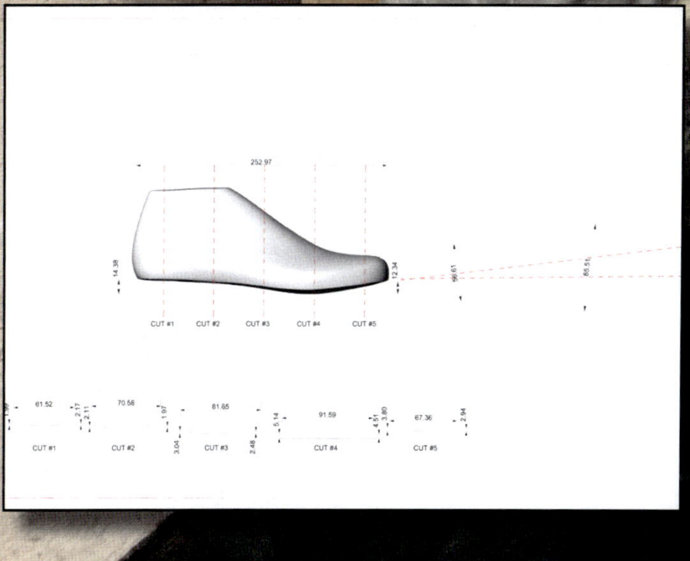

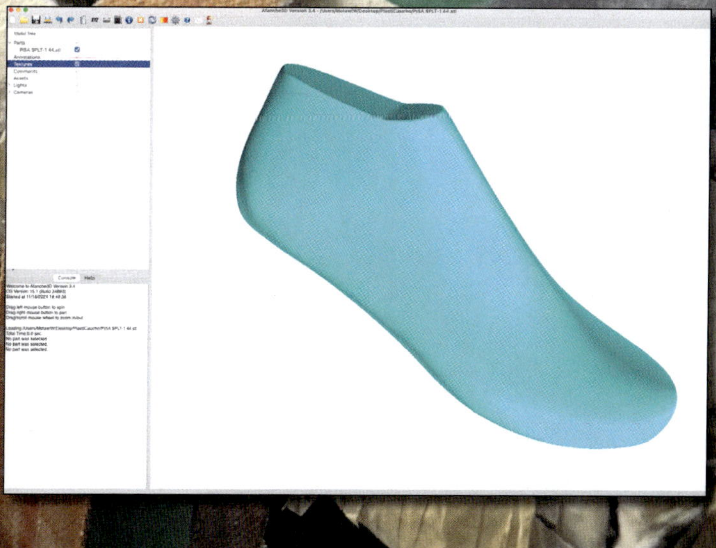

Types of Shoe Lasts and Materials

The choice of material for a shoe last depends on the manufacturing process:

Plastic Lasts: Used primarily for modern cold-cement construction, plastic lasts start as large injection-molded blocks that are then machined to precise specifications.

Metal Lasts: Required for vulcanized and injected footwear, metal lasts—typically cast aluminum—are necessary due to their durability and ability to withstand high temperatures. Some manufacturers now produce plastic lasts capable of enduring the vulcanization process.

Automation and CNC-Machined Lasts:
In large-scale production, CNC cut lasts made from metal or advanced composites are used to ensure precision and consistency. These are particularly common in high-efficiency manufacturing environments.

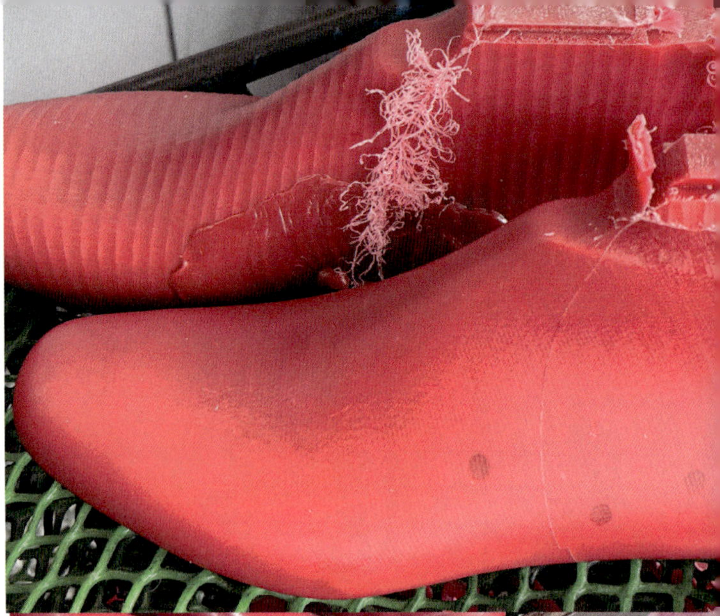

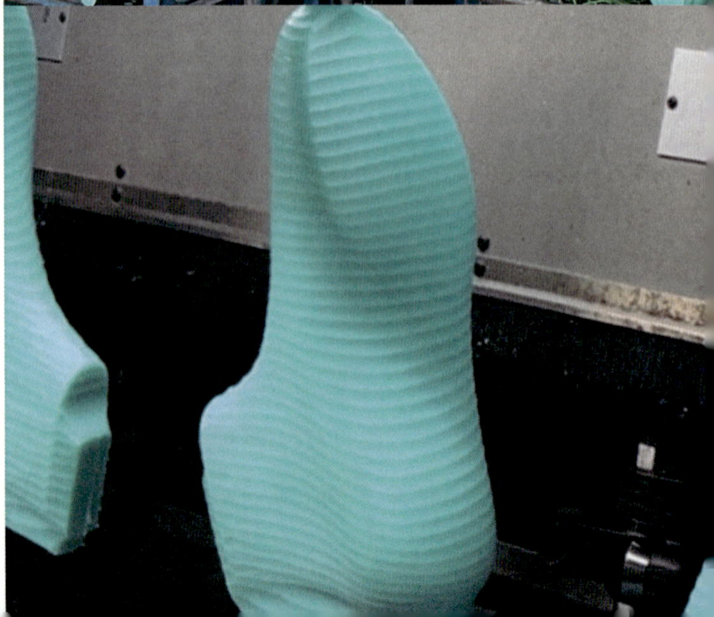

Solid metal lasts

Articulating metal lasts

Cost and Investment Considerations

The cost of shoe lasts varies depending on material, complexity, and production volume. In bulk, a solid plastic last may cost between $12 and $15, while a hinged, articulated last can range from $15 to $20. For custom development, a shoe factory may charge approximately $500 per last.

The number of lasts required depends on production volume. A small brand placing an order for 1,000 to 1,500 pairs may require 200 to 220 lasts.

In high-volume factories, each assembly line may need 750 to 800 lasts, and multiple lines running simultaneously can significantly increase this requirement. Large brands placing substantial orders may not be charged separately for additional lasts, as factories may absorb the cost due to the scale of production.

Metal lasts for injection

Automatic machines require CNC cut precision lasts. Depending the design these can cost from $200 to $100 per size.

Production Workflow and Last Storage

Efficient last storage and management are essential for streamlined production. In well-organized factories, lasts are stored by size to ensure smooth workflow integration. Once shoes move through the assembly line, the lasts are retrieved via conveyor belts for continuous reuse. Proper storage and organization prevent damage and ensure longevity.

Conclusion

The shoe last is a fundamental element in footwear manufacturing, directly influencing the design, fit, and functionality of a shoe. It is essential to invest in high-quality, properly designed lasts, as they serve as the foundation for all subsequent tooling and production processes.

Proper last development and management enhance efficiency, ensure product consistency, and contribute to the overall success of a footwear brand. By understanding the significance of lasts and making informed choices, manufacturers can optimize their production processes and create high-quality footwear that meets both aesthetic and functional requirements.

Chapter 5

Creating Your Footwear Pattern

Once the shoe last has been established, the pattern can be created. The pattern represents the design of the shoe and consists of multiple parts, even for a simple shoe.

The Pattern

What is the pattern?
The outline shape of the cut parts. Every shoe has a pattern .

A simple sneaker pattern

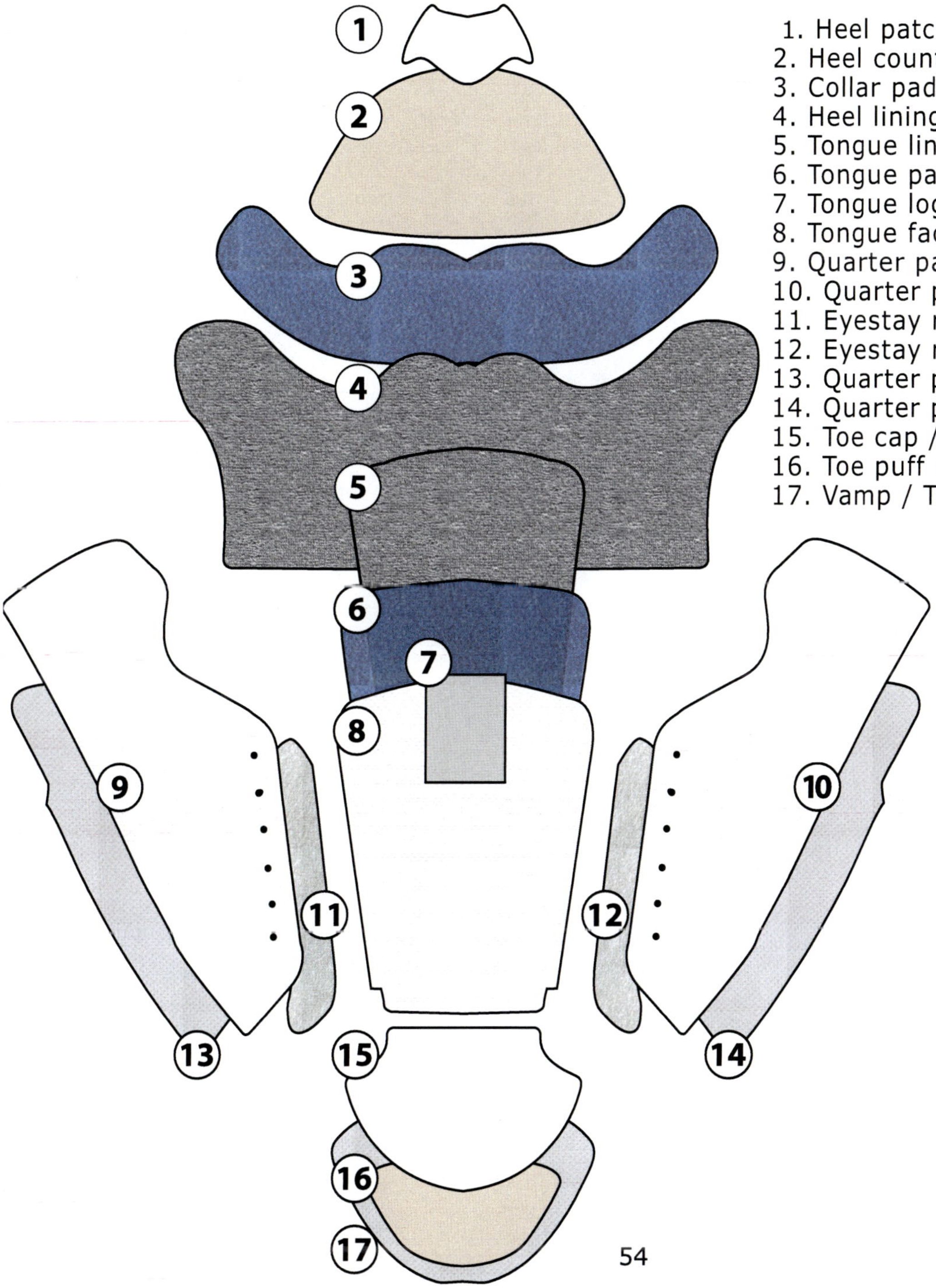

1. Heel patch or Mustache
2. Heel counter reinforcement
3. Collar padding
4. Heel lining
5. Tongue lining
6. Tongue padding
7. Tongue logo panel
8. Tongue face
9. Quarter panel medial side
10. Quarter panel lateral side
11. Eyestay reinforcement medial
12. Eyestay reinforcement lateral
13. Quarter panel lining medial side
14. Quarter panel lining lateral side
15. Toe cap / Vamp
16. Toe puff reinforcement
17. Vamp / Toe cap lining

Pattern Making

Pattern making is crucial because it defines the components required for shoe construction and influences the shape of any injection or upper mold pieces. A standard sneaker pattern may have up to 17 different parts, illustrating the complexity of the process.

The traditional approach involves taping the last, marking it, and flattening it to create a pattern. This method is still commonly used in factories, although digital processes are becoming increasingly prevalent.

Computer-aided design (CAD) systems allow for the creation of precise flat patterns, often starting with a paper pattern that is scanned into the system.

These digital patterns display essential design lines, including details such as the tongue, the lasting apron, and various structural elements.

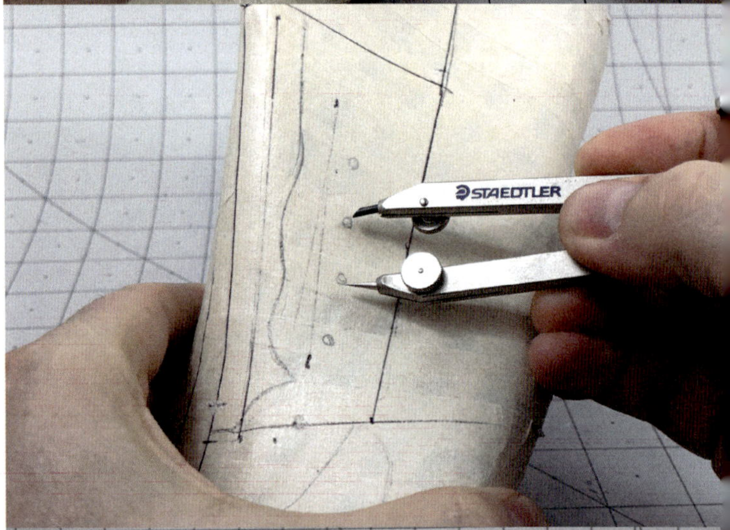

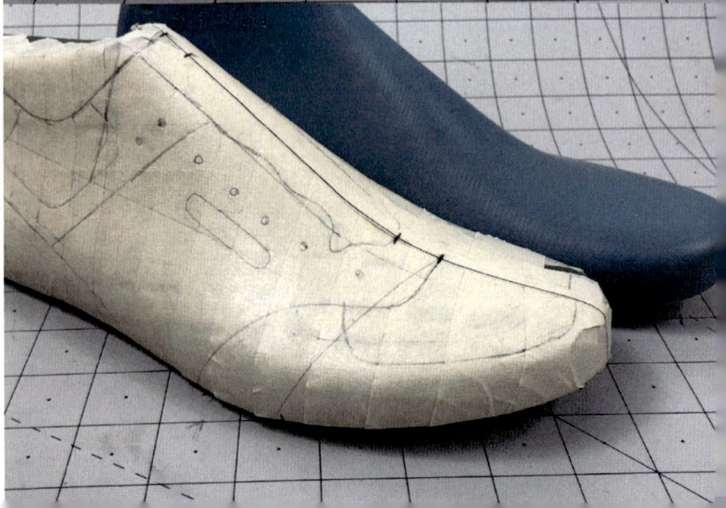

Pattern Testing

Once the pattern is finalized, it is cut into paper or plastic templates, which serve as guides for sample production. The sample department uses these patterns to hand-cut materials and assemble test shoes. Pattern makers play a critical role in this process, making necessary adjustments to ensure accuracy.

While designers and developers may not be experts in pattern making, they must communicate specific construction requirements to pattern makers, ensuring that the final design aligns with the intended specifications.

After the pattern has been finalized, it can be digitized to create a precise flat pattern for cutting components. These components are then assembled into a prototype, often referred to as a "pullover," which is a preliminary version of the upper that has been shaped around the last. A computer-generated design may not always perfectly match the pullover, but this discrepancy is part of the development process. The pullover is then scanned to determine the final shape of the outsole, factoring in the thickness of the upper materials.

With the pattern and sample upper completed, the next step is the production of cutting dies. Cutting dies are essential for mass production as they allow materials to be efficiently cut to the required shapes using hydraulic presses. Cutting dies come in various forms, such as those for footbeds, toe tips, and vamps. They feature sharp edges to ensure precision cutting.

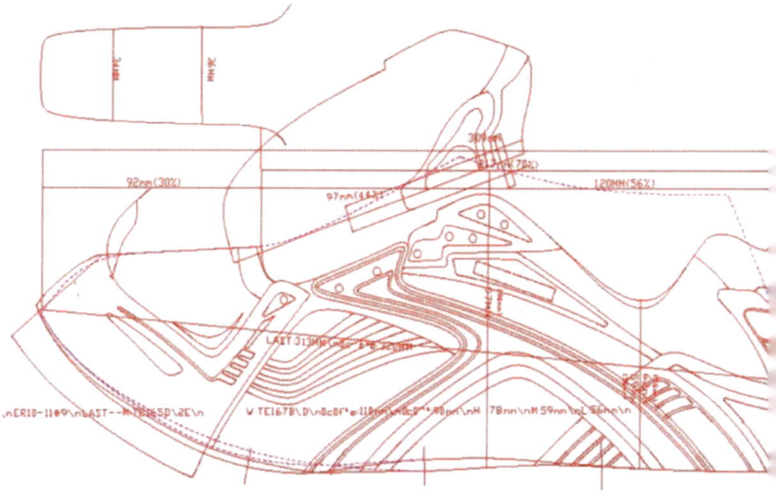

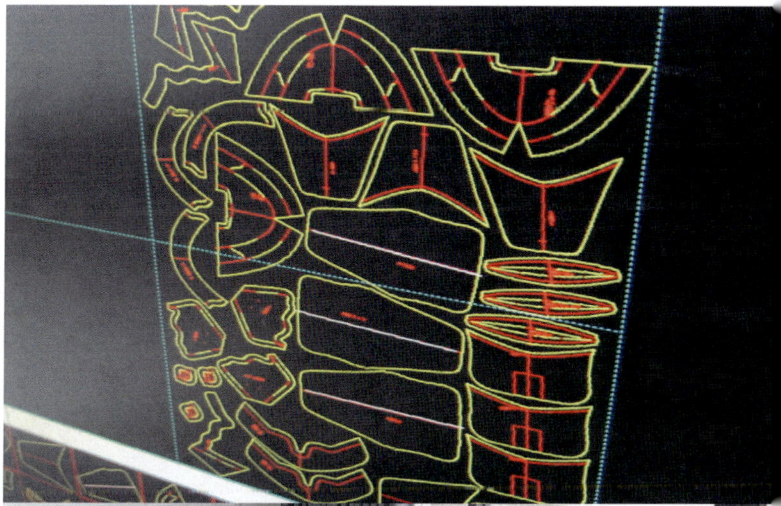

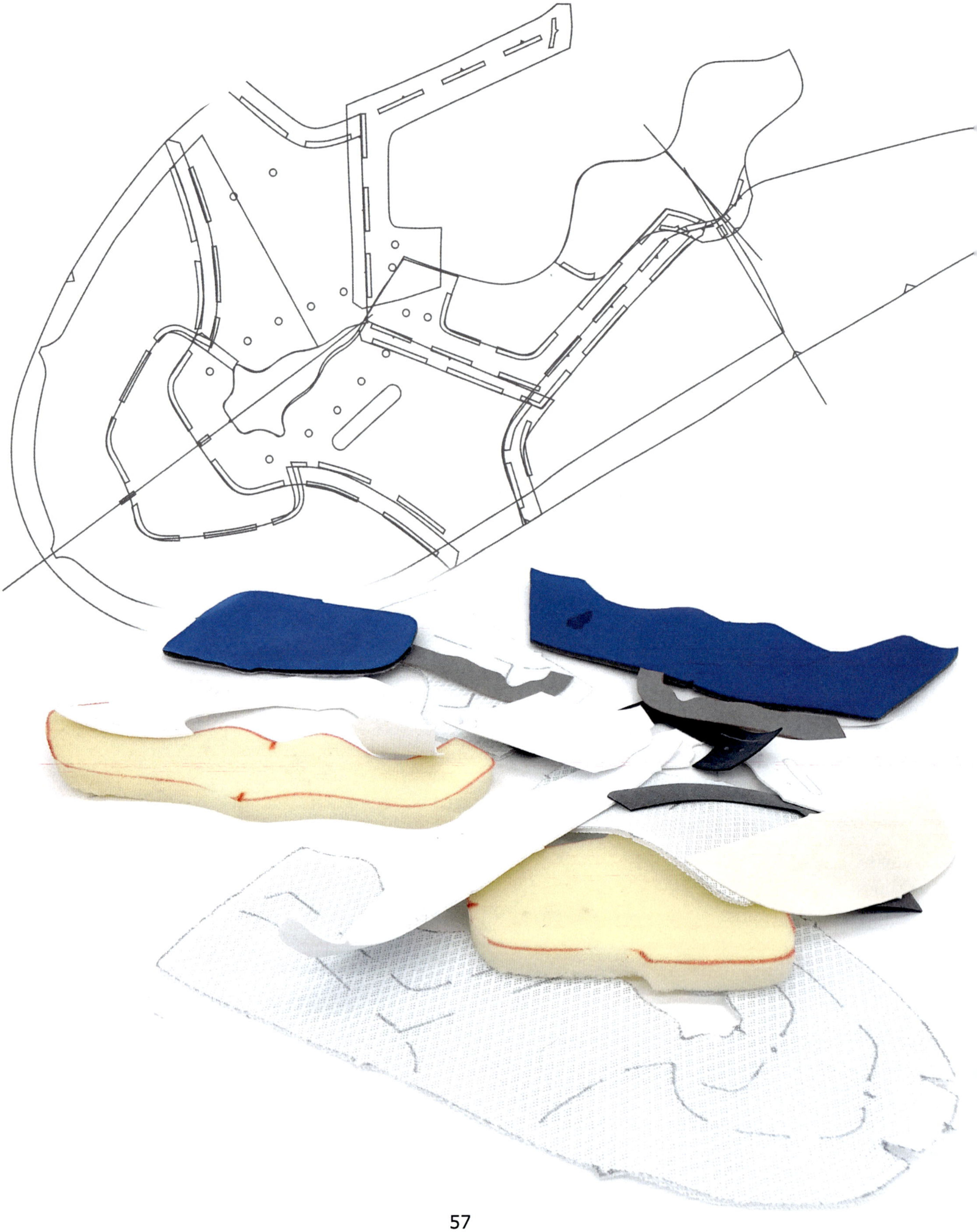

Once you have the pattern and pull-over you can get serious about making your cutting dies and outsole tooling.

Concept rendering

Pullover prototype also called a "lasted upper"

Chapter 6

Making Footwear Cutting Dies

With the pattern and sample upper completed, the next step is the production of cutting dies.

Cutting dies are essential for mass production as they allow materials to be efficiently cut to the required shapes using hydraulic presses.

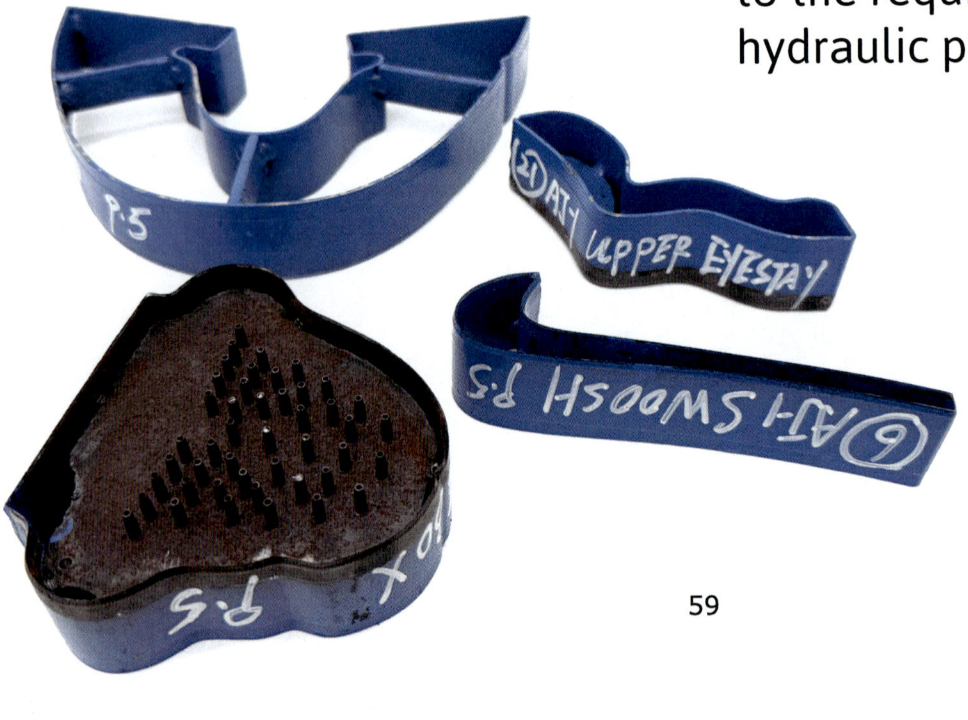

4

Making Cutting Dies

Cutting dies are not manufactured within the shoe factory itself. Instead, the factory sends pattern specifications to a specialized workshop where the dies are created. Each piece of the shoe requires its own cutting die, and each size variation necessitates a separate set of dies.

While automated cutting methods are becoming more common, traditional cutting dies remain the preferred choice for leather footwear.

Hydraulic presses are commonly used for die-cutting, requiring operators to manually insert materials and activate the press using a dual-button safety mechanism.

Efficient cutting minimizes material waste, as any excess material must either be discarded or recycled.

Cutting dies have various features, including perforations and alignment marks. The manufacturing process begins with straight steel pieces that are bent into shape using specialized machinery.

Skilled technicians shape and assemble the dies based on design specifications. Once a die is formed, it is welded for durability, with cross-bracing added for additional strength.

There are different types of cutting dies. The most basic type consists of a bent metal blade, which arrives pre-sharpened and cannot be resharpened once worn out.

More advanced versions include wood-framed or plexiglass-supported dies, which are used for applications requiring high precision and visibility. Laser-cut frames may be used to ensure accurate alignment.

CNC machines are employed to manufacture highly intricate cutting dies, particularly for detailed logo elements.

When a cutting die becomes obsolete—whether due to wear, rust, or design changes—it is discarded and replaced. The lifespan of a cutting die depends on production volume and material hardness.

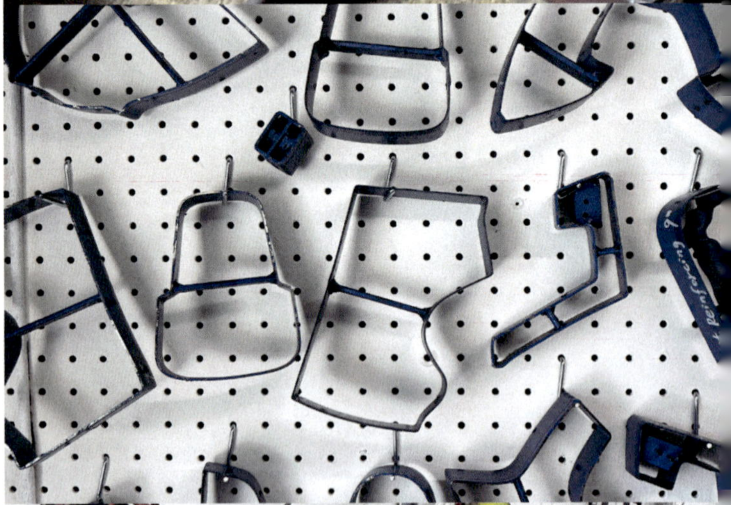

A common hydraulic cutting press

Die makers follow
the paper pattern

Die bending stations

鞋身里内。9.5
DC24220
鞋身里内。10
DC24220
DC24220

Steel blades with plastic base for precision cutting

Installing perforation pins

Size marking
notches

Centerline
markings

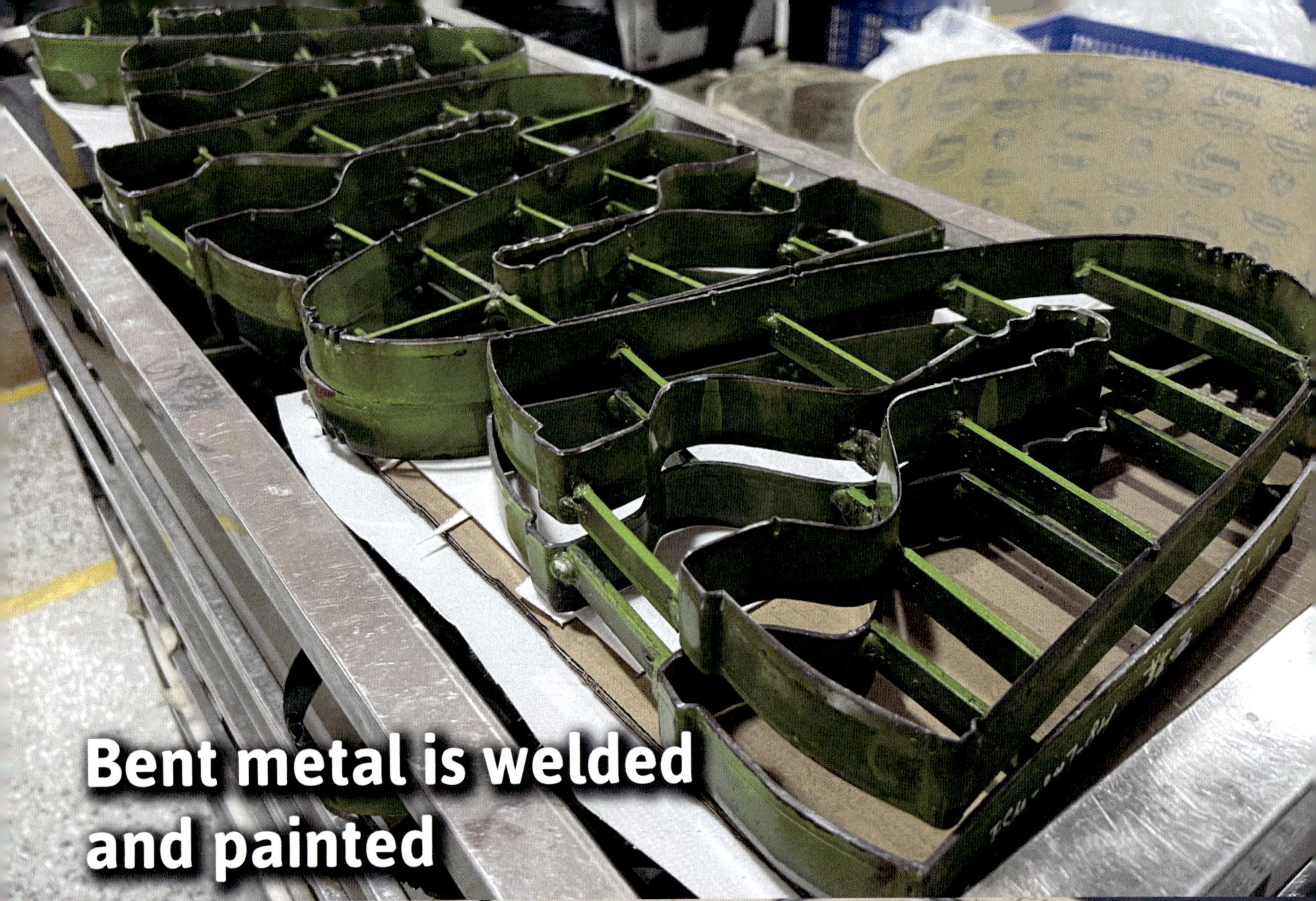

Bent metal is welded and painted

Wood framed dies with locating pins

Plastic frames
with see through
openings

CNC dies for
precision cutting

Cut parts with size notches

Features on cutting dies

Size marks
It's almost impossible to determine the size of a cut shoe component. The difference between a size 7 vamp and a 7.5 vamp may be only a millimeter or two. You can see on the edges of these cut components a system of notches which clearly shows the sizes. When the parts are stacked, it's easy to see if you have parts that are all the same size.

Centering marks
The cutting die will also be marked with the center point to help the stitchers align the parts correctly.

Alignment guides
Factory pattern makers may add pattern assembly guides or nibs to the pattern. As a designer, I find these nibs unsightly and try to remove them if possible. These nibs are also used by authenticators to determine if a product is real or not.

There is usually a workstation behind the cutting press. As the cutter works, a co-worker is organizing, counting, and checking all the parts and usually using stickers to label the bundles. This operation is critical to keep the parts organized for the stitching department.

Can you imagine hundreds of unmarked parts? How could the stitching department possibly assemble them correctly?

Cut parts are counted, stacked organized and inventoried

A full set of cutting dies ready for production

How much does the pattern cost?
Expect to pay $300 or $400 dollars for the first pattern. Repeat customers or high volume projects may be $0.

How much are cutting dies?
Typical steel rule dies cost around $150 per size.
Laser framed and plastic framed dies cost more per part but you need fewer of these.

CNC cutting dies will be a few hundred dollars per set but you may need just one or two per size.

How many do I need?
You will need one set per each size. Large production runs will require more dies but you may not see a bill for these.

Cutting dies are consumable
Footwear cutting dies are consumable. It's difficult or impossible to repair or re-sharpen cutting dies. If your orders are large enough to wear out some cutting dies the shoe factory will likely replace them without sending you a bill.

Spent cutting dies will be recycled for their steel or reused to decorate the shoe factory. This type of welded cutting defense is not uncommon.

Chapter 7

Outsole Tooling Development Process

With the last complete and pattern created it's time to make the outsole.

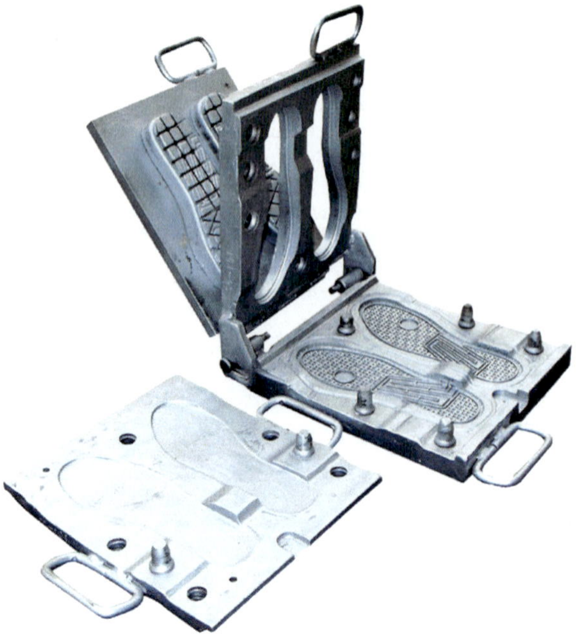

Scanning operations
Introduction of 3D scanning technology revolutionized the footwear tooling production process. Before scanning, technicians would measure the lasted upper and record the contours with a manual profile gauge.

Below is an example.

The 3D optical scanner has replaced this tedious and error prone process.

You can see the black and white scan dots all over the lasted upper in the photo below. The scan dots help the computer recognize the surface contour of the object to be scanned.

Scan the completed pull overs

Drawing The Parts

Drawing Your Parts
With design drawings and the 3D scan in hand the tooling factory technician can get started drawing the parts. The first step is to draw the parts, the mold comes later.

Using the 3D scans, the technician can make sure that the upper and the outer will precisely fit together. Remember, this is a challenge with open molds. If the parts weren't designed to fit together, they will NEVER fit together correctly. The scanning technology is also used to make sure that any rubber and EVA parts that are designed to be bonded together do, in fact, fit together.

The same 3D scan data can also be used in conjunction with a pattern to create any upper molds that have 3D shaping. Any flat molds for flexible parts do not require the upper scan data. 3D components such as heel counters or toe caps will require a 3D mold.

The 3D scanning and modeling process done in the factory ensures the upper and the outsole will fit together. This task is the responsibility of the shoe factory.

Technicians get started drawing your parts

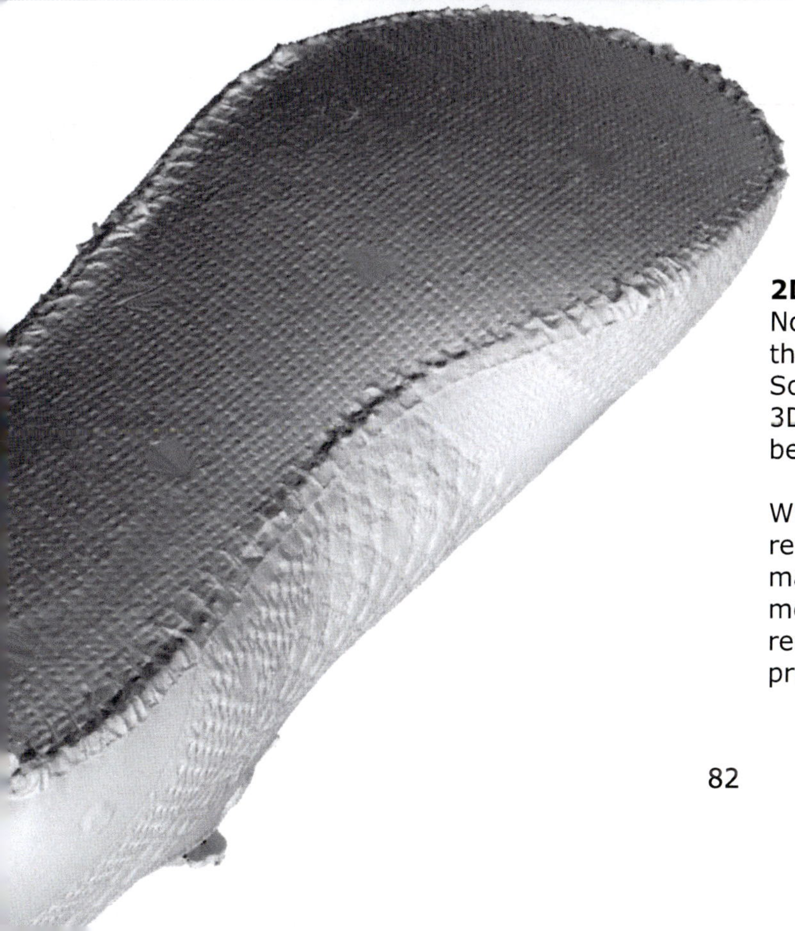

2D vs 3D drawings
Now that we have 3D technology you may think that a 2D drawing is obsolete. This is not true. Some factories will be content to supply only a 3D drawing and you must demand a 2D blueprint before they create the 3D drawing.

While the 3D drawing is an excellent tool to review the cosmetic design and ensure it matches the shoe designer's intent, the 3D model does not reveal all of the cross-sections required to ensure the design is engineered properly.

Reading & Reviewing Tooling Drawings

Learning how to read a blueprint is an important skill you will need to master. I've been drawing and reading blueprints for over 40 years.
My advice to you is to go slow and study the drawing methodically. Spend at least 30 minutes with a drawing even if every detail looks right. I guarantee you are missing something.

Ddon't worry about sending back the drawing for revisions. At this stage in the process, we are spending time not money.

The blueprint is the record of your design decisions and the instructions for your mold maker. The small details are important. Once you approve the drawing someone is going to be cutting metal.

TOP VIEW
SECTION A-A'
SECTION E-E'
LATERAL VIEW
SECTION D-D'
BOTTOM VIEW
SECTION C-C'
MEDIAL VIEW
内侧图
PRINTING
TPU

Outsole Terminology

Before proceeding, it is essential to understand the technical terms related to mold drawings.

Knowing the correct terminology will facilitate effective communication and ensure accuracy in the manufacturing process.

1. **Lugs**
2. **Web**
3. **Sections**
4. **Bonding Margin**
5. **Top Net**
6. **Toe Drop**
7. **Stack Height**
8. **Toe Spring**
9. **Heel Entry**
10. **Parting Lines**
11. **Color Dams**

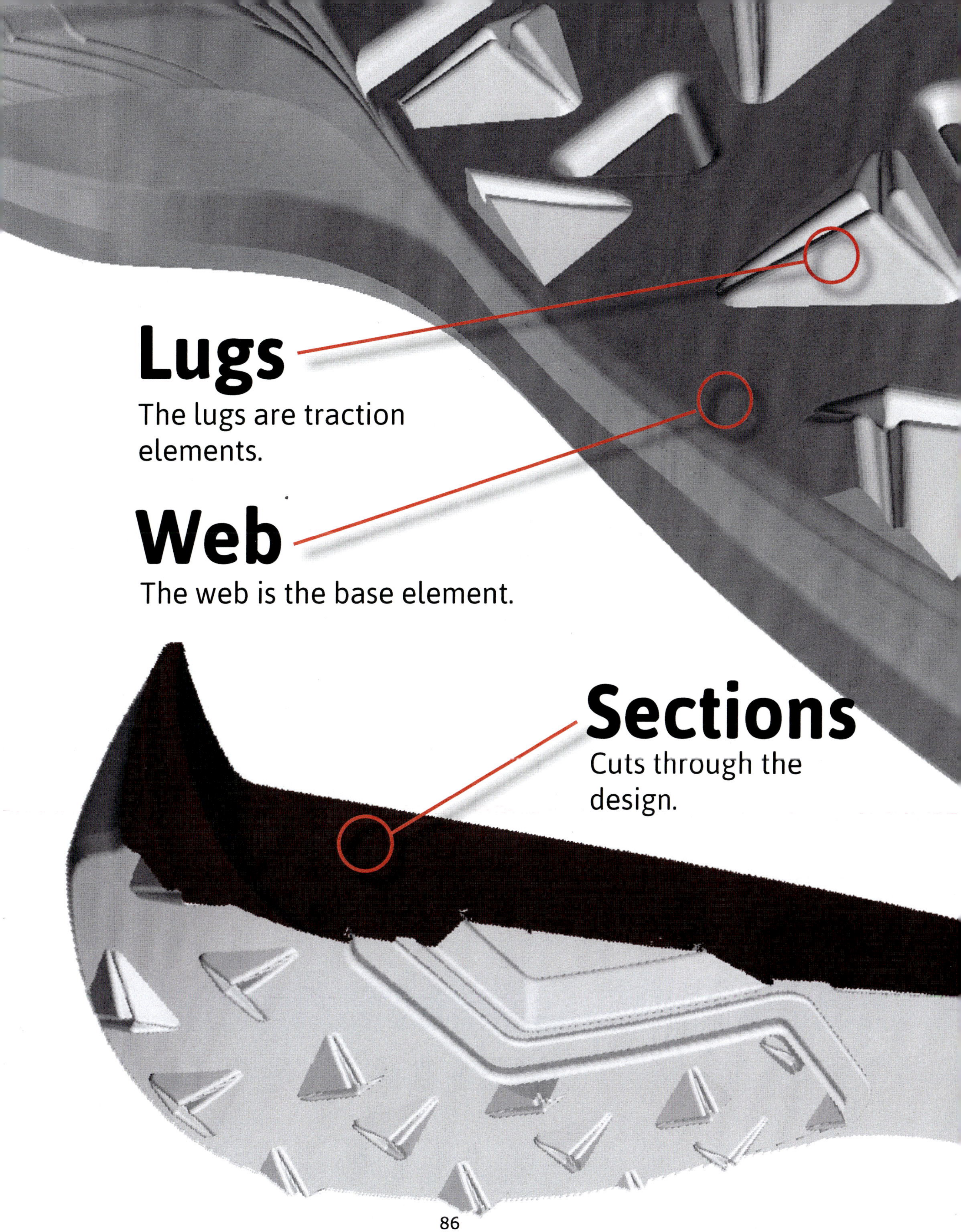
Lugs
The lugs are traction elements.
Web
The web is the base element.
Sections
Cuts through the design.

Top Net

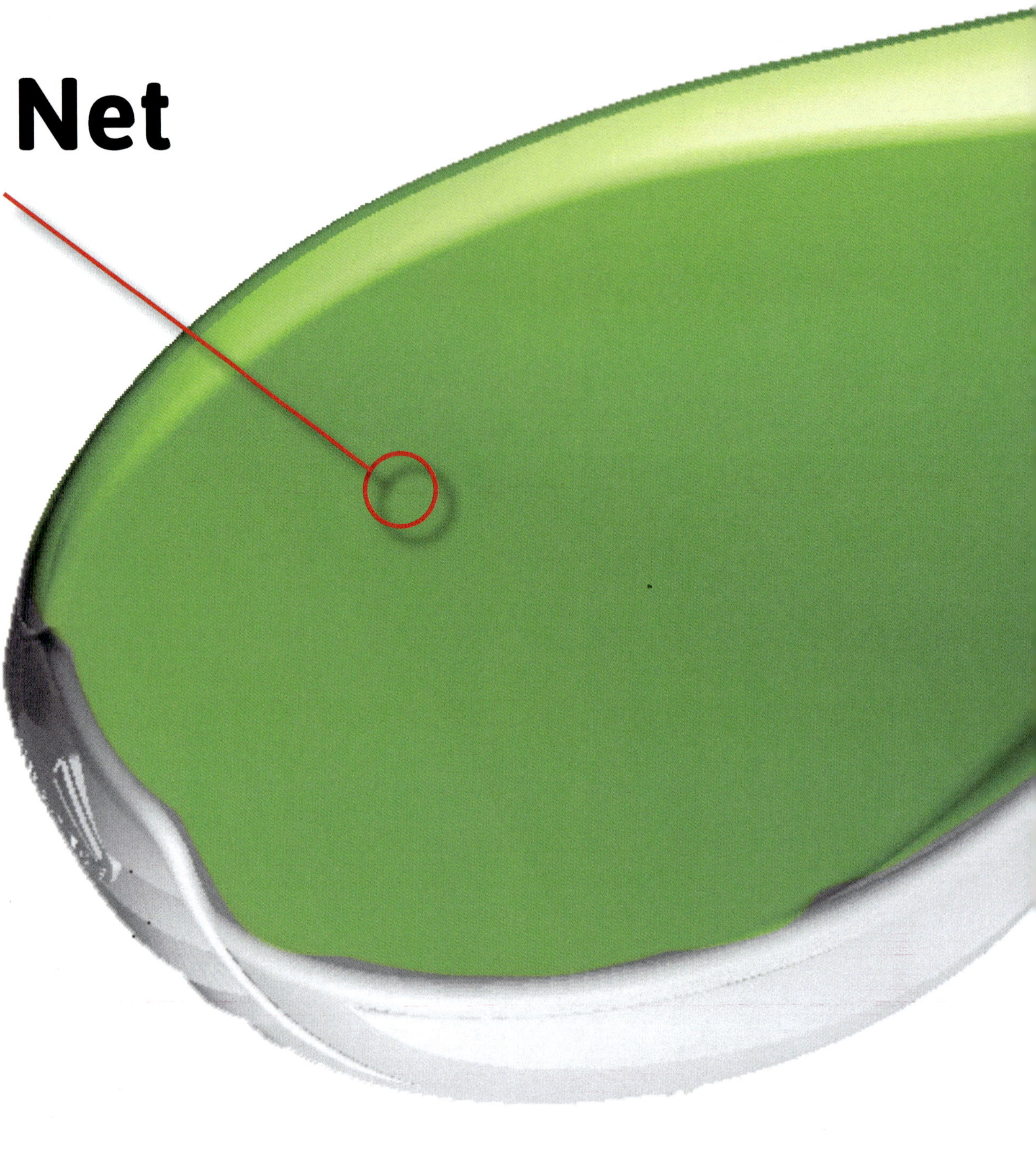

Top Net – This is the shape that matches the bottom of the lasted upper. A properly designed top net will ensure the outsole fits correctly.

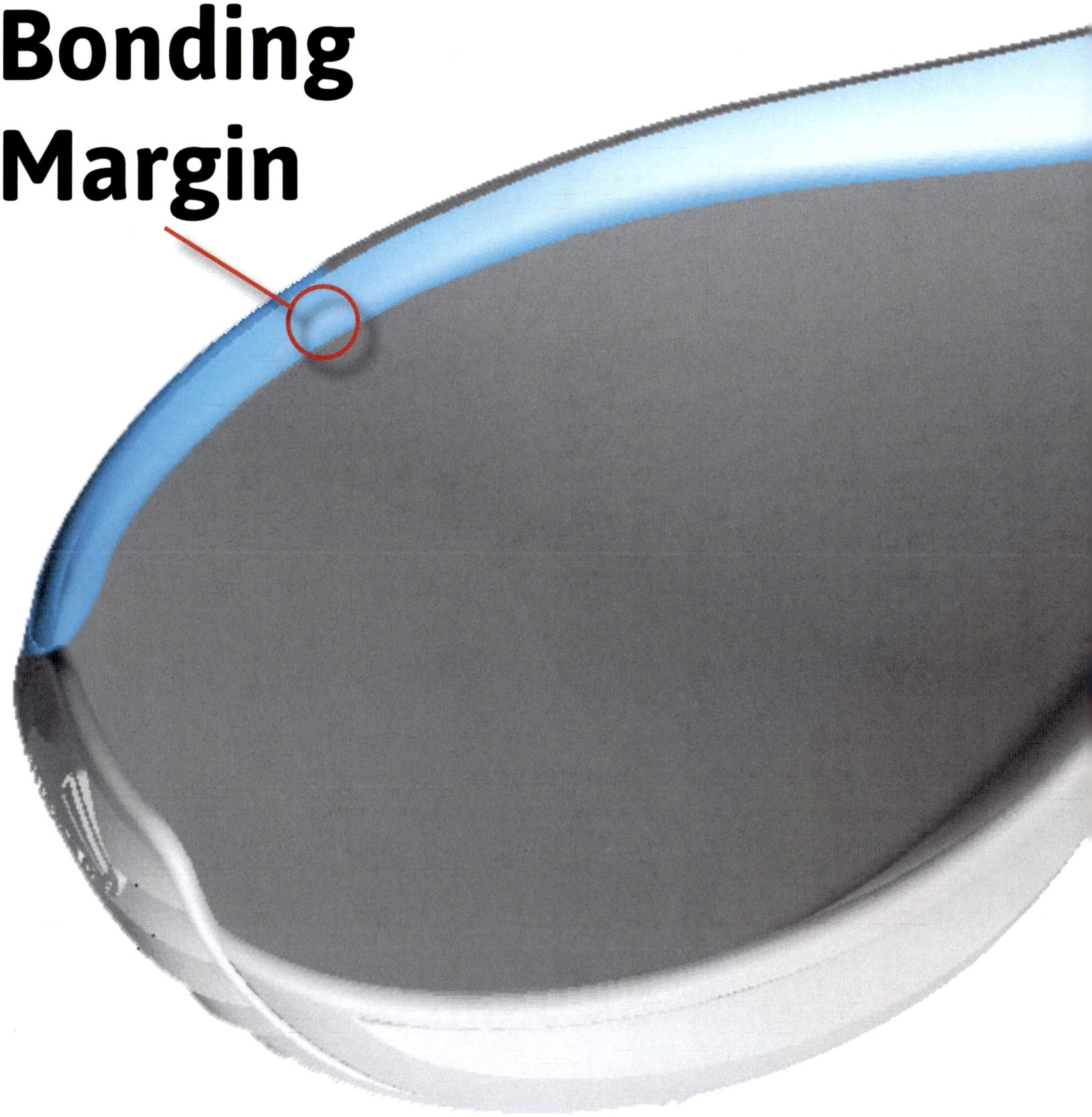

Bonding Margin – The vertical wall around the perimeter that ensures strong adhesion. A minimum of 5mm is recommended and 8mm is ideal.

Bonding Margin

Bonding Margin

Flat bonds are good

No Problem

Toe Drop

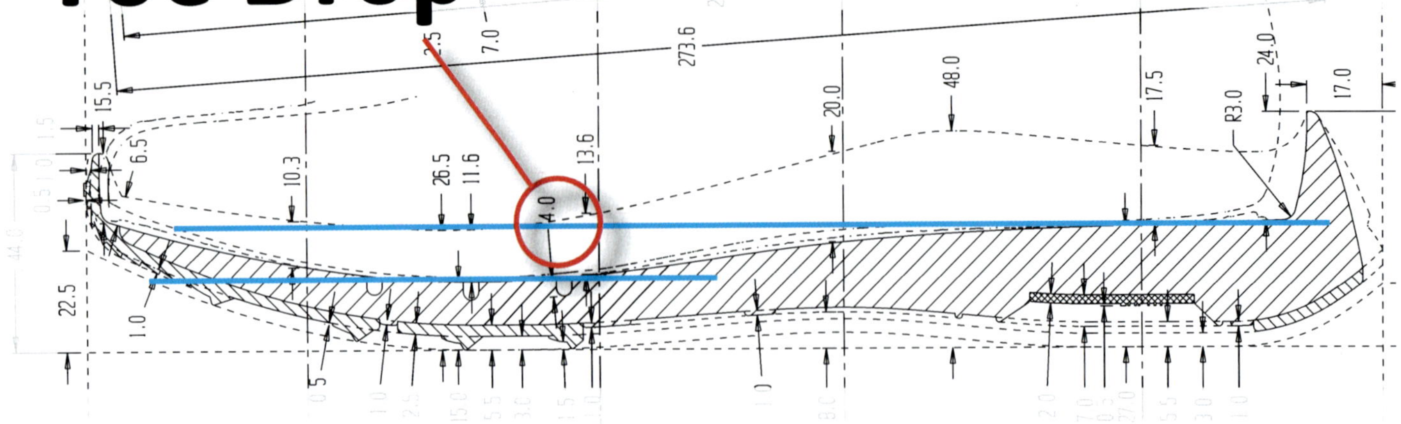

The toe drop is built into the last...NOT the outsole.

Stack Height

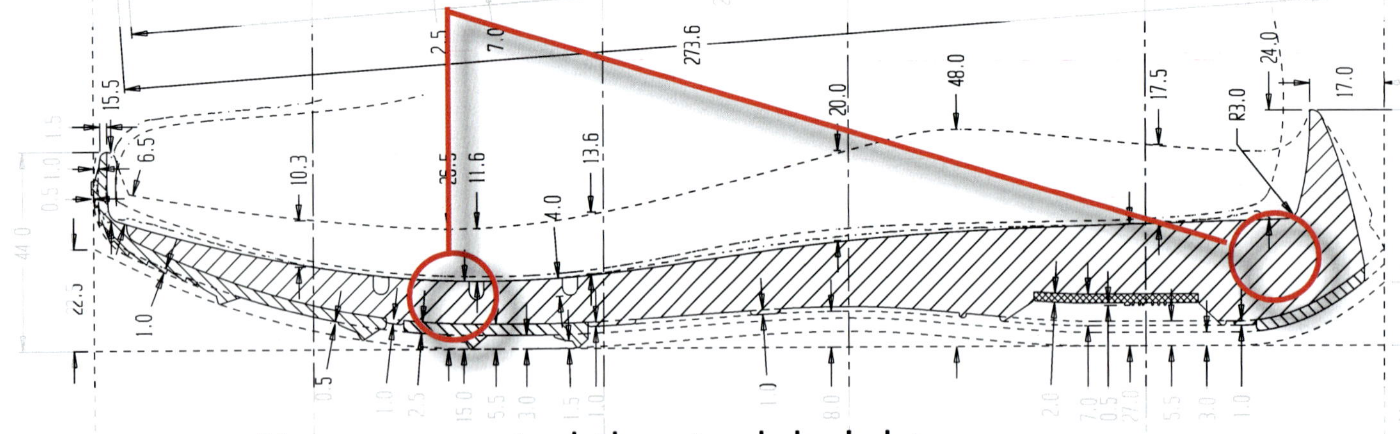

You can control the stack height to add in more cushioning.

Toe spring is controlled by the last bottom contour and the midsole thickness.

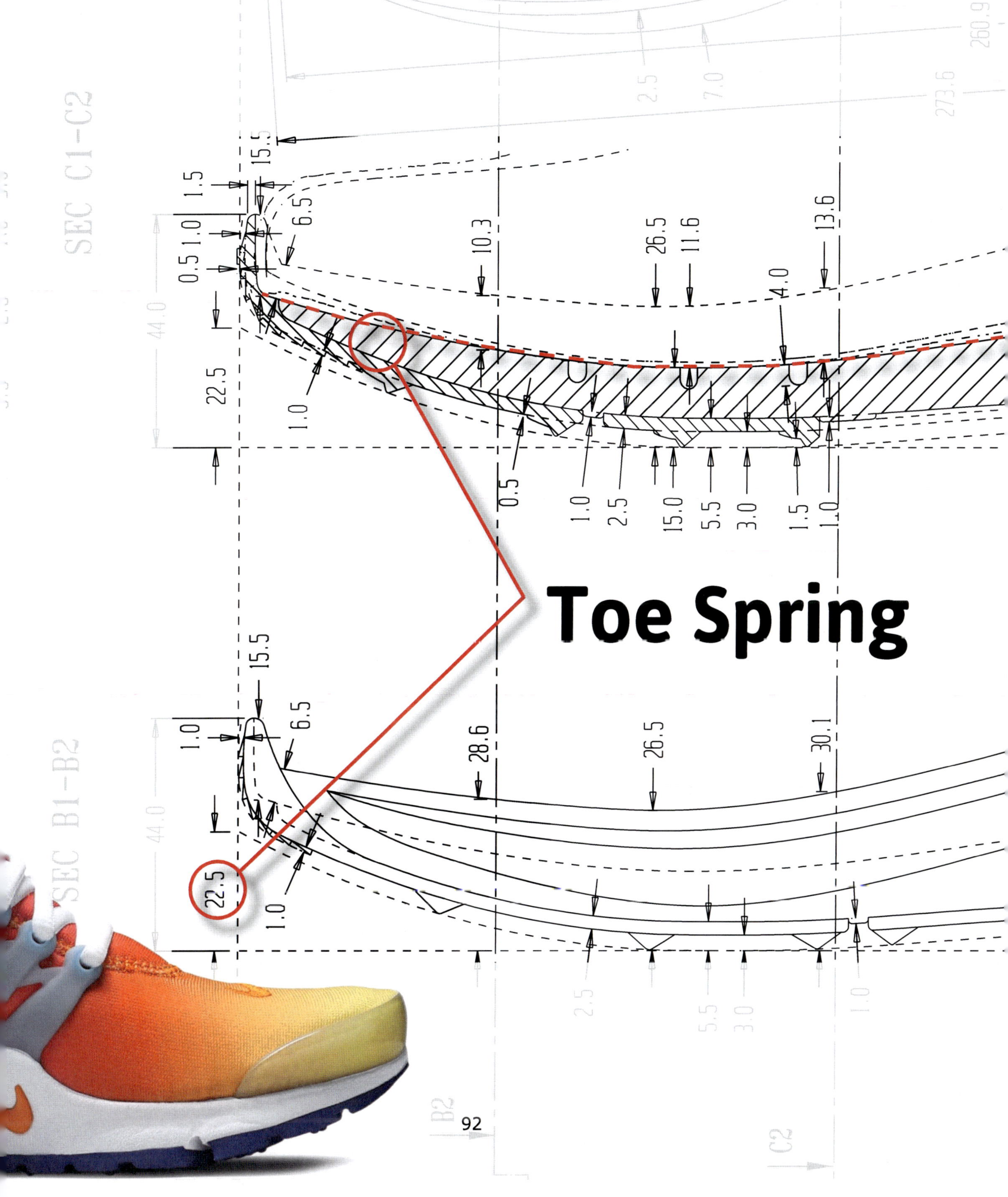

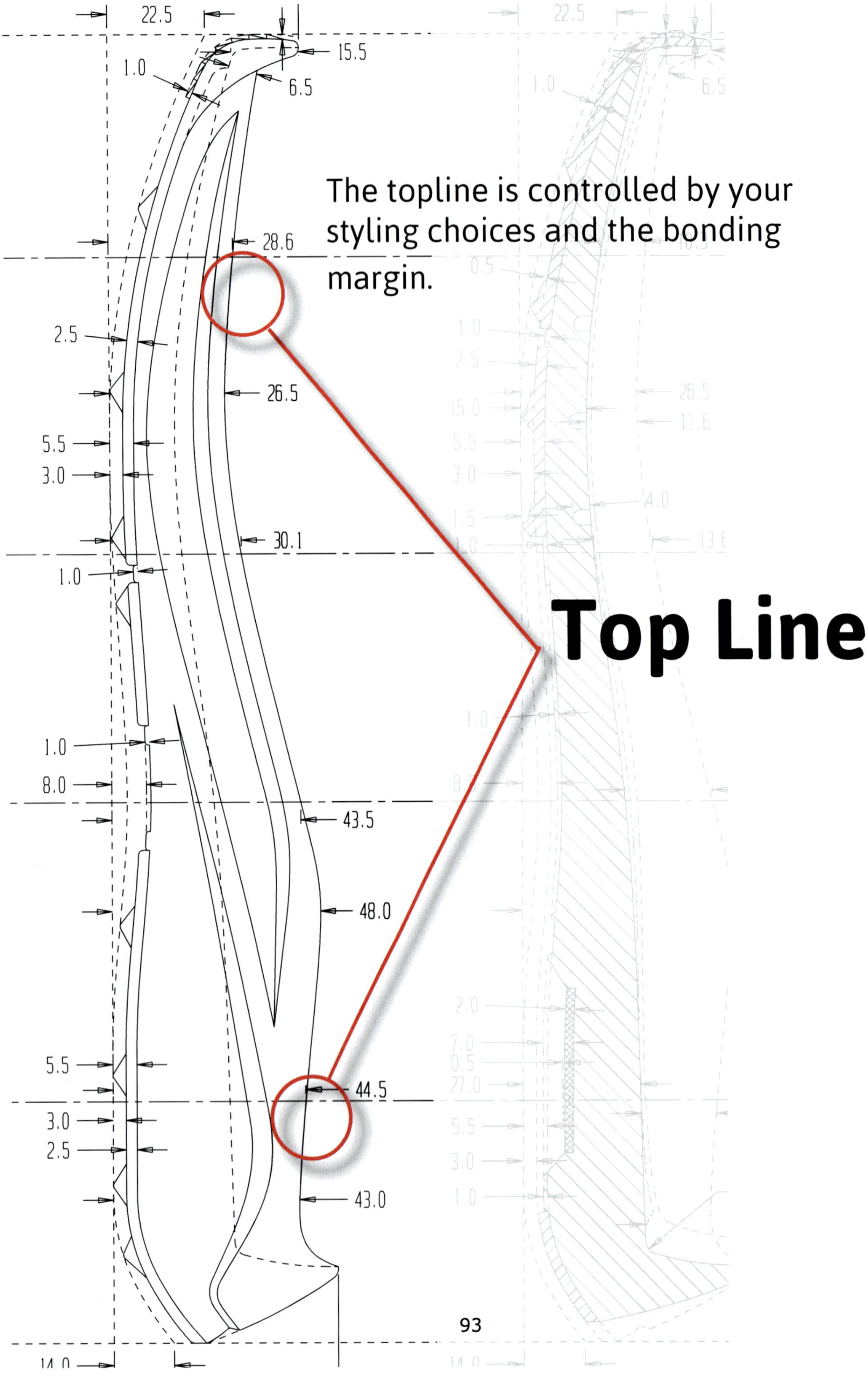

The topline is controlled by your styling choices and the bonding margin.

Heel Strike Entry Angle

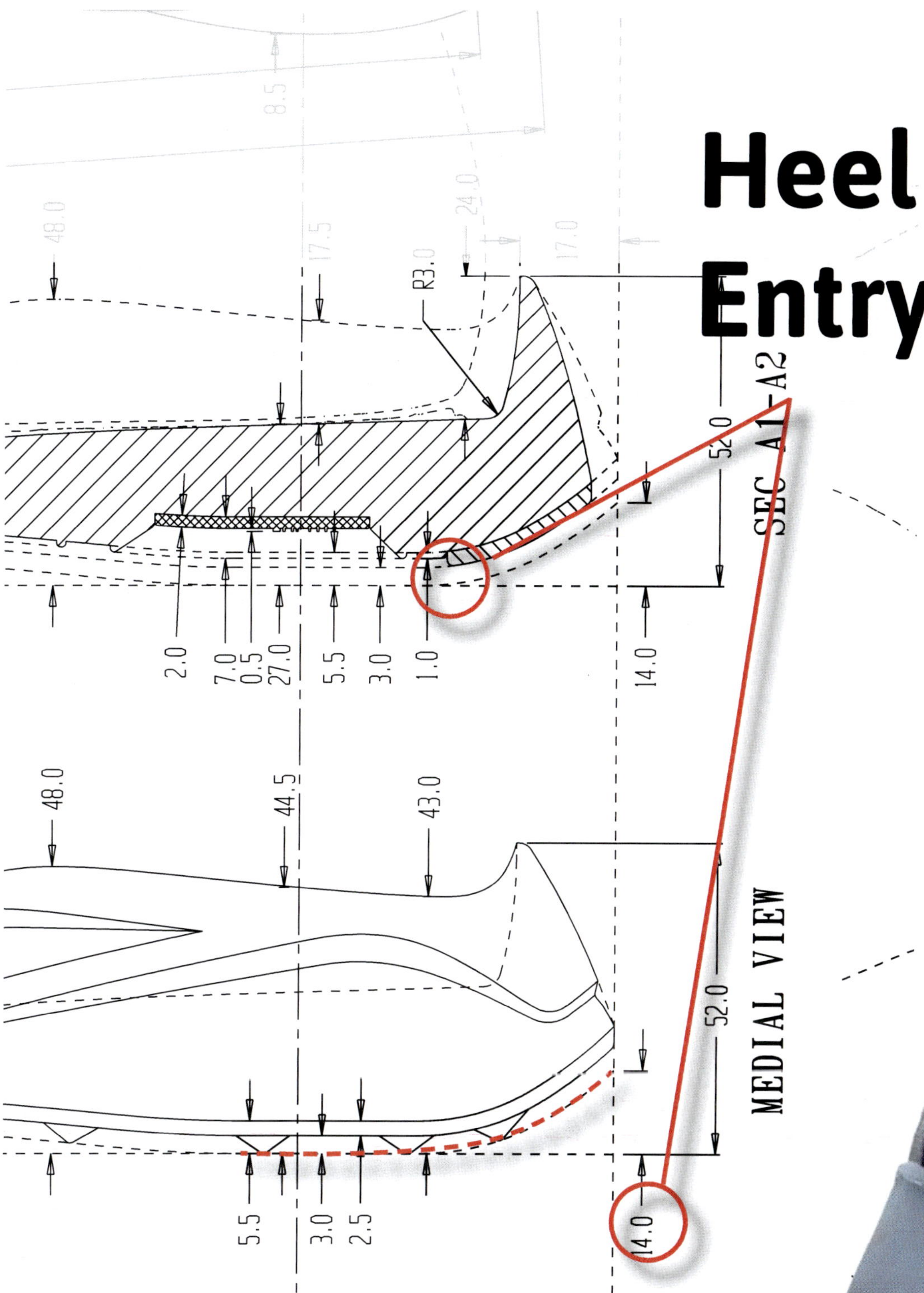

For runners, the heel strike and entry angle are fine points they will notice instantly.

Color Dams

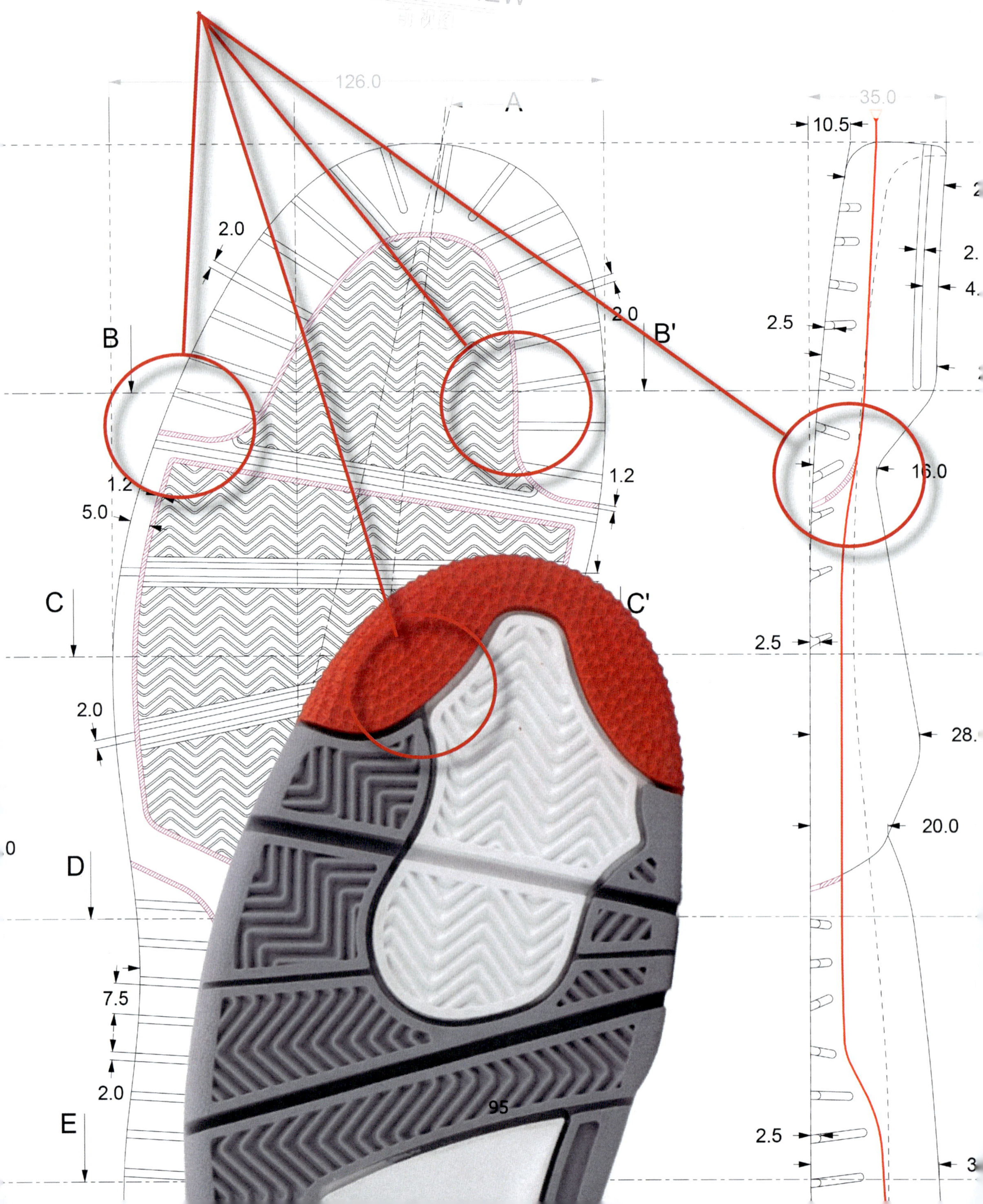
FRONT VIEW
35.0
126.0
A
35.0
10.5
2.0
2.0
B
B'
2.5
16.0
1.2
1.2
5.0
C
C'
2.5
2.0
28.
20.0
D
7.5
2.0
E
2.5

Color dams and mold textures

Parting Lines

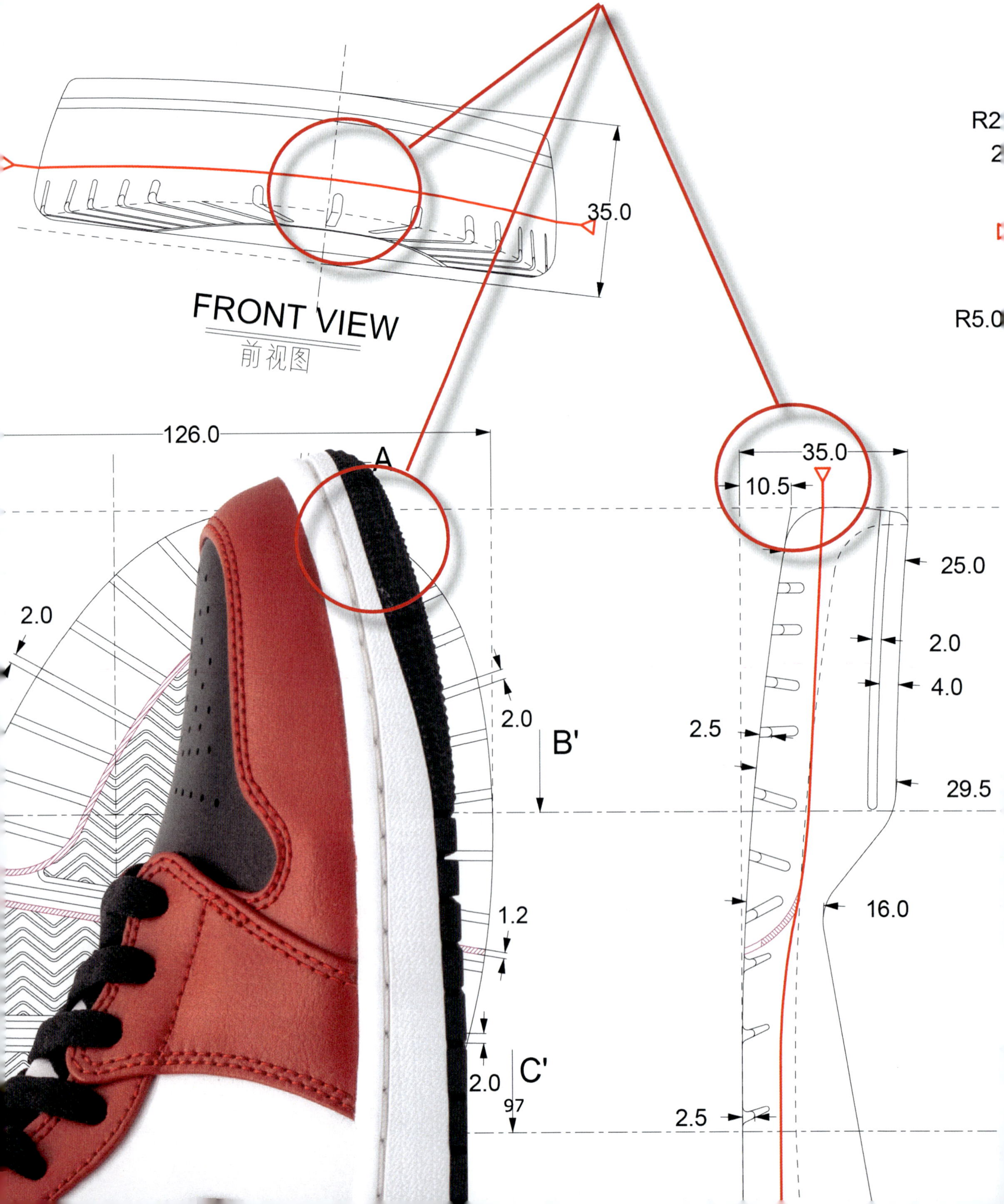

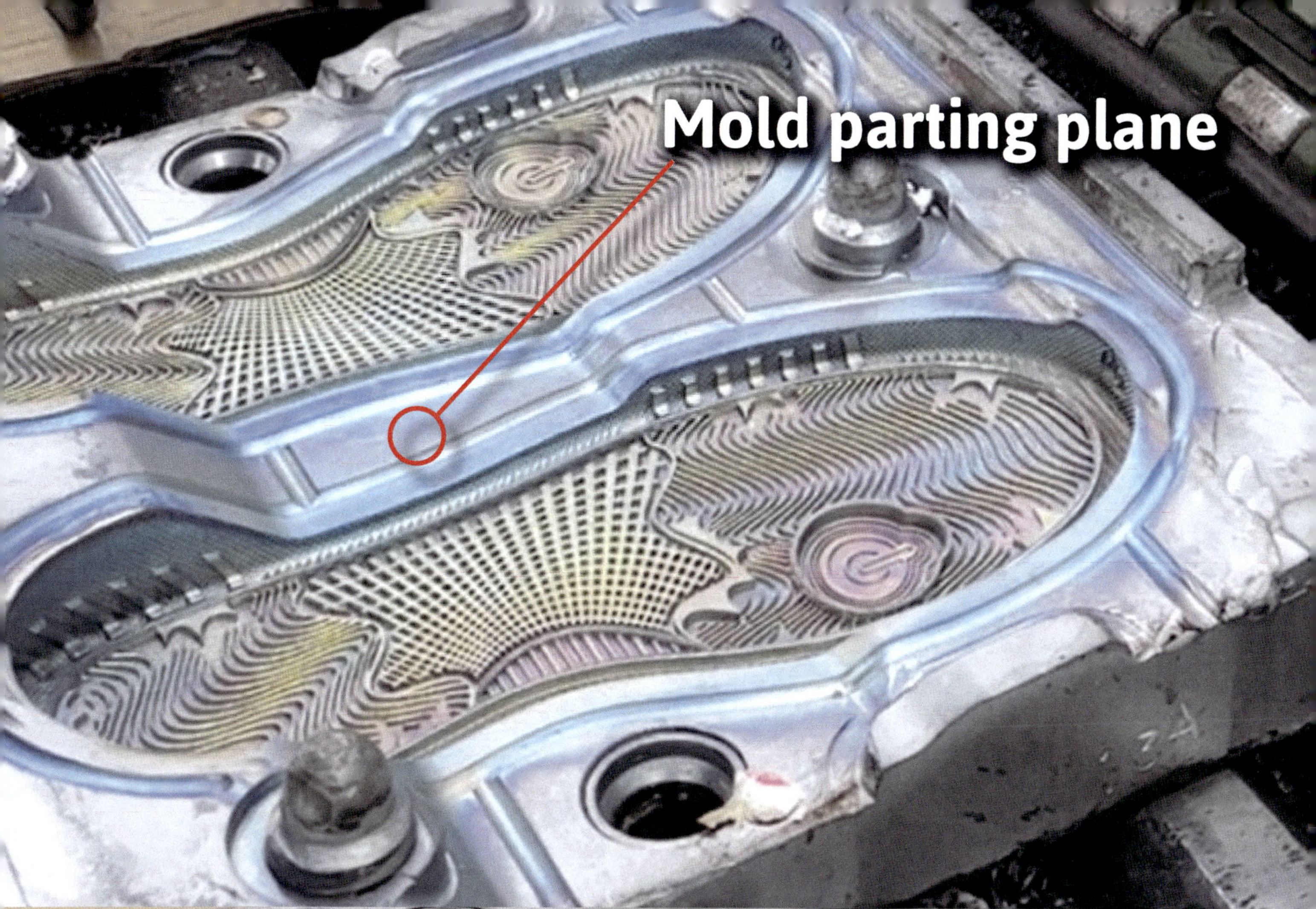
Mold parting plane

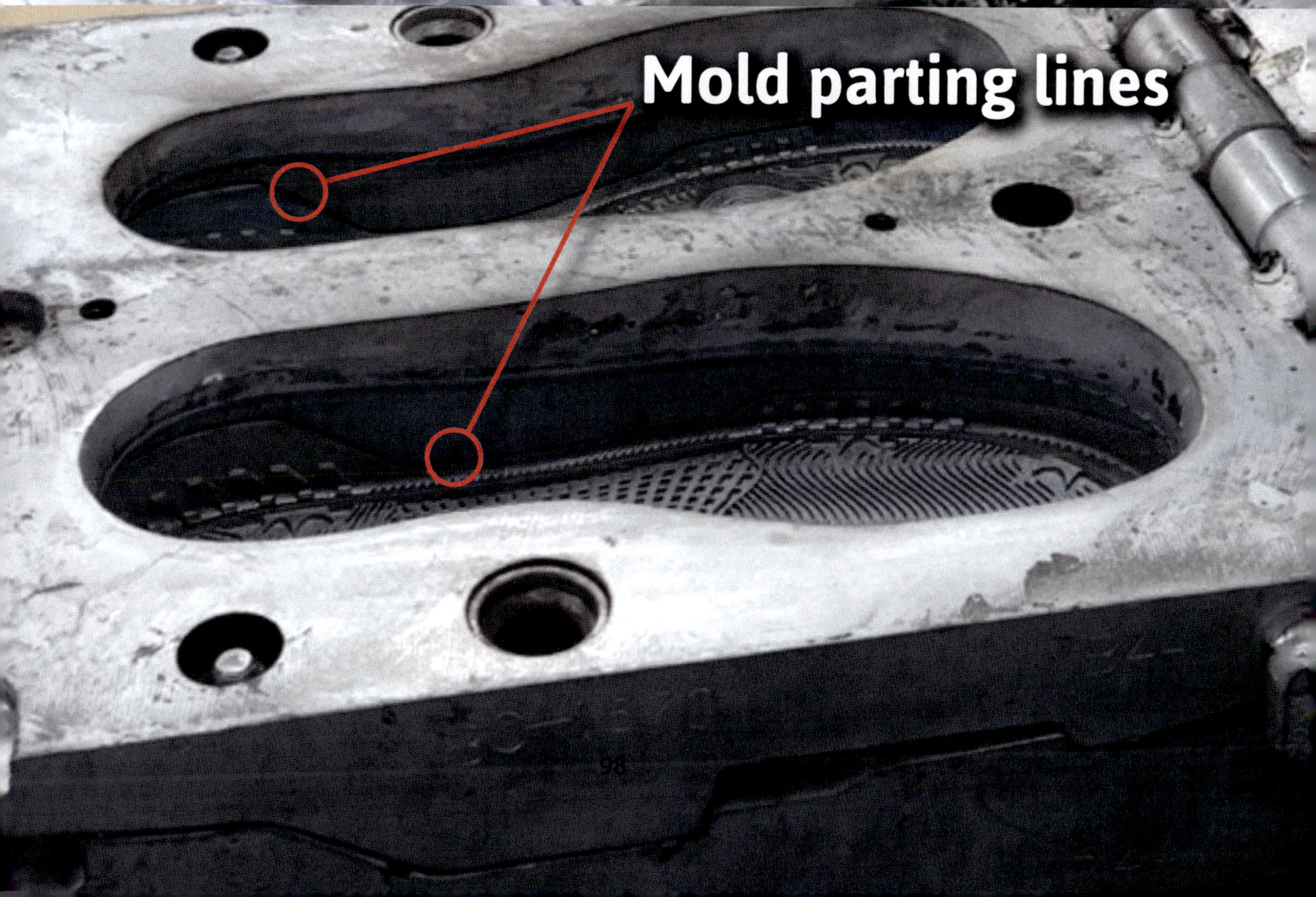
Mold parting lines

The 50 Point Blueprint Review Checklist

Mold Review 50 Point Check List

Look At The Title Block:

1. The item model name
2. The mold shop name and customer name
3. The shoe last code number
4. Drawing part size
5. The material call out
6. Revision list
7. Component hardness specs or color information

On The Drawing You Should See:

8. Bottom view tread
9. Lateral side view (outsole)
10. Medial side view (inside)
11. Toe view
12. Heel view
13. Top net - top surface (top net) view of the midsole
14. Last bottom contour line
15. Rubber top line
16. Midsole top line
17. Cross section A-A: toe to heel
18. Forefoot section B-B
19. Arch section C-C
20. Waist section D-D
21. Heel section E-E

Review All Critical Measurements:

22. Web thickness
23. Lug heights
24. Toe spring
25. Forefoot midsole thickness
26. Heel midsole thickness
27. Heel strike entry height
28. Heel strike length
29. Top net corner radius
30. Toe drop
31. Cross sections detailing the lug pattern

Check The Design Details:

32. Is the texture map complete?
33. Evaluate the weight - study the rubber cross sections
34. Look at the last bottom contours in all sections
35. Flex grooves - review on the top net and bottom rubber
36. Logo size, depth, orientation
37. Mold parting line locations
38. Bonding margins for the upper and any sub components
39. Bonding margins for subcomponents - outsole to midsole
40. Review cross sections of any technology items
41. Shank - location, thickness, & bonding margins
42. Air bags - location, thickness, & bonding margins
43. Gel inserts - location, thickness & bonding margins
44. Side wall details - Do they match the designers intent?
45. Compare left and right shoe logo orientations
46. Review high-wear areas - forefoot and heel strike
47. Windows - penetrating the midsole or rubber parts?
48. Parting lines for color breaks
49. Color dams
50. Is the part design suitable for the specified materials?

SHOEMAKERS ACADEMY

Let's go!

Gather your checklist, blue prints and some measuring tools.

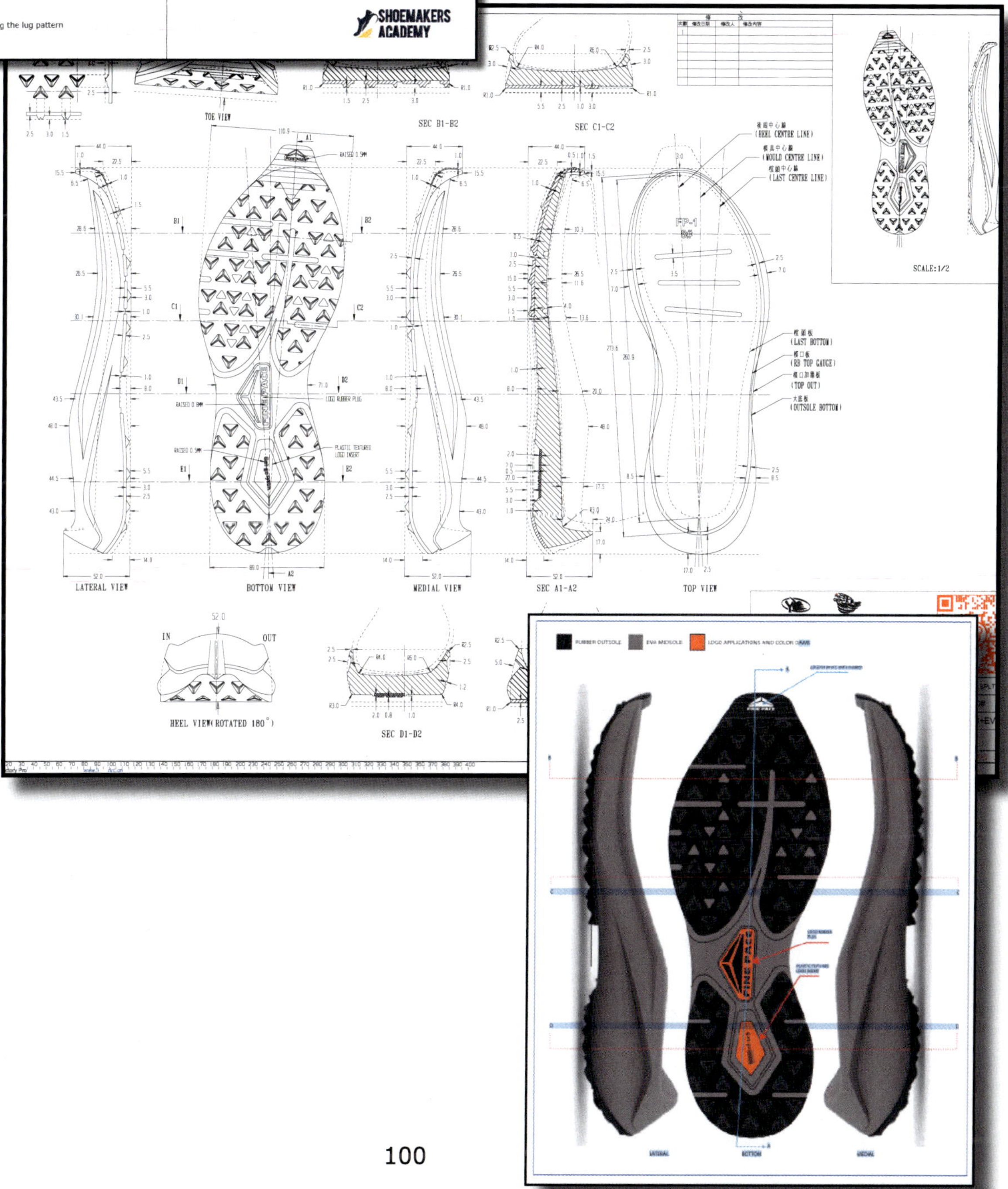

Mold Review: 50 Point Checklist

Look At The Title Block:

1 The item model name
2 The mold shop name and customer name
3 The shoe last code number
4 Drawing part size
5 The material call out
6 Revision list
7 Component hardness specs or color information

On The Drawing You Should See:

8 Bottom view tread
9 Lateral side view (outsole)
10 Medial side view (inside)
11 Toe view
12 Heel view
13 Top net - top surface (top net) view of the midsole
14 Last bottom contour line
15 Rubber top line
16 Midsole top line
17 Cross section A-A: toe to heel
18 Forefoot section B-B
19 Arch section C-C
20 Waist section D-D
21 Heel section E-E

Review All Critical Measurements:

22 Web thickness
23 Lug heights
24 Toe spring
25 Forefoot midsole thickness
26 Heel midsole thickness
27 Heel strike entry height
28 Heel strike length
29 Top net corner radius
30 Toe drop
31 Cross sections detailing the lug pattern

Check The Design Details:

32 Is the texture map complete?
33 Evaluate the weight - study the rubber cross sections
34 Look at the last bottom contours in all sections
35 Flex grooves - review on the top net and bottom rubber
36 Logo size, depth, orientation
37 Mold parting line locations
38 Bonding margins for the upper and any subcomponents
39 Bonding margins for subcomponents - outsole to midsole
40 Review cross sections of any technology items
41 Shank - location, thickness, & bonding margins
42 Airbags - location, thickness, & bonding margins
43 Gel inserts - location, thickness & bonding margins
44 Side wall details - Do they match the designer's intent?
45 Compare left and right shoe logo orientations
46 Review high-wear areas - forefoot and heel strike
47 Windows - penetrating the midsole or rubber parts?
48 Parting lines for color breaks
49 Color dams
50 Is the part design suitable for the specified materials?

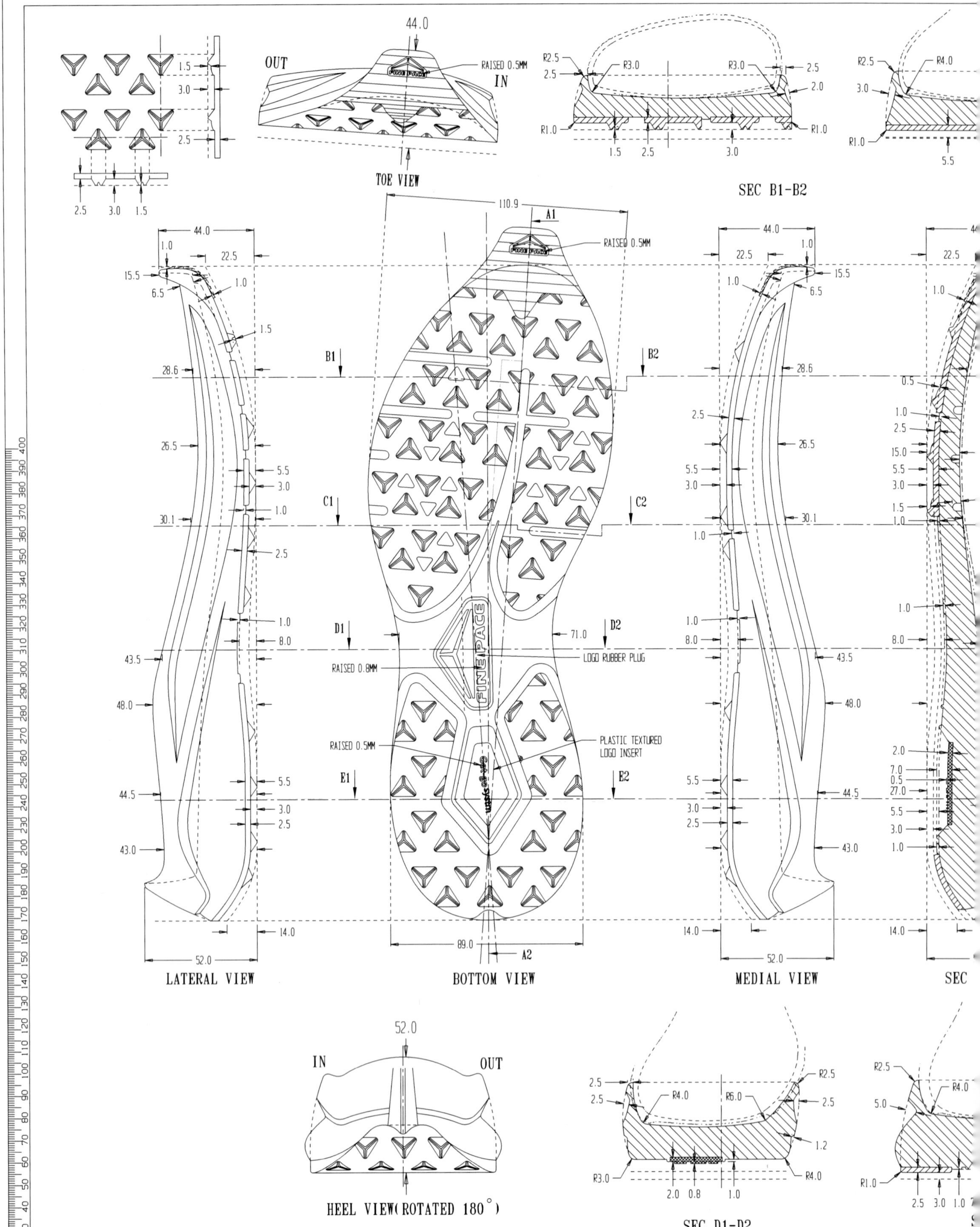

OUT
IN
RAISED 0.5MM
TOE VIEW
SEC B1-B2
110.9
A1
B1
B2
C1
C2
D1
D2
E1
E2
A2
89.0
71.0
LOGO RUBBER PLUG
RAISED 0.8MM
PLASTIC TEXTURED
LOGO INSERT
FINE PACE
LATERAL VIEW
BOTTOM VIEW
MEDIAL VIEW
SEC
HEEL VIEW(ROTATED 180°)
SEC D1-D2

修	改		
次數	修改日期	修改人	修改内容
1			

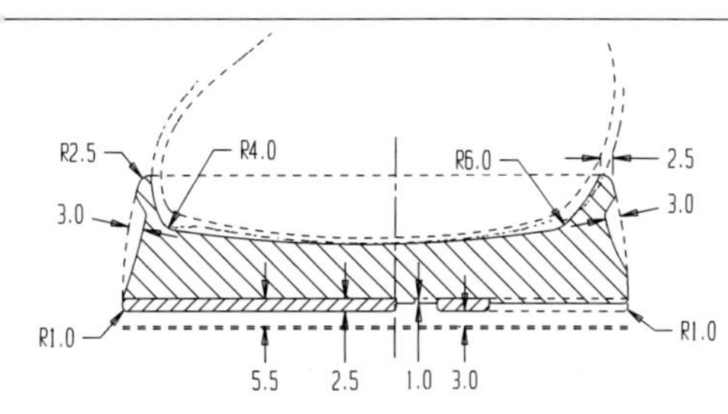

SEC C1-C2

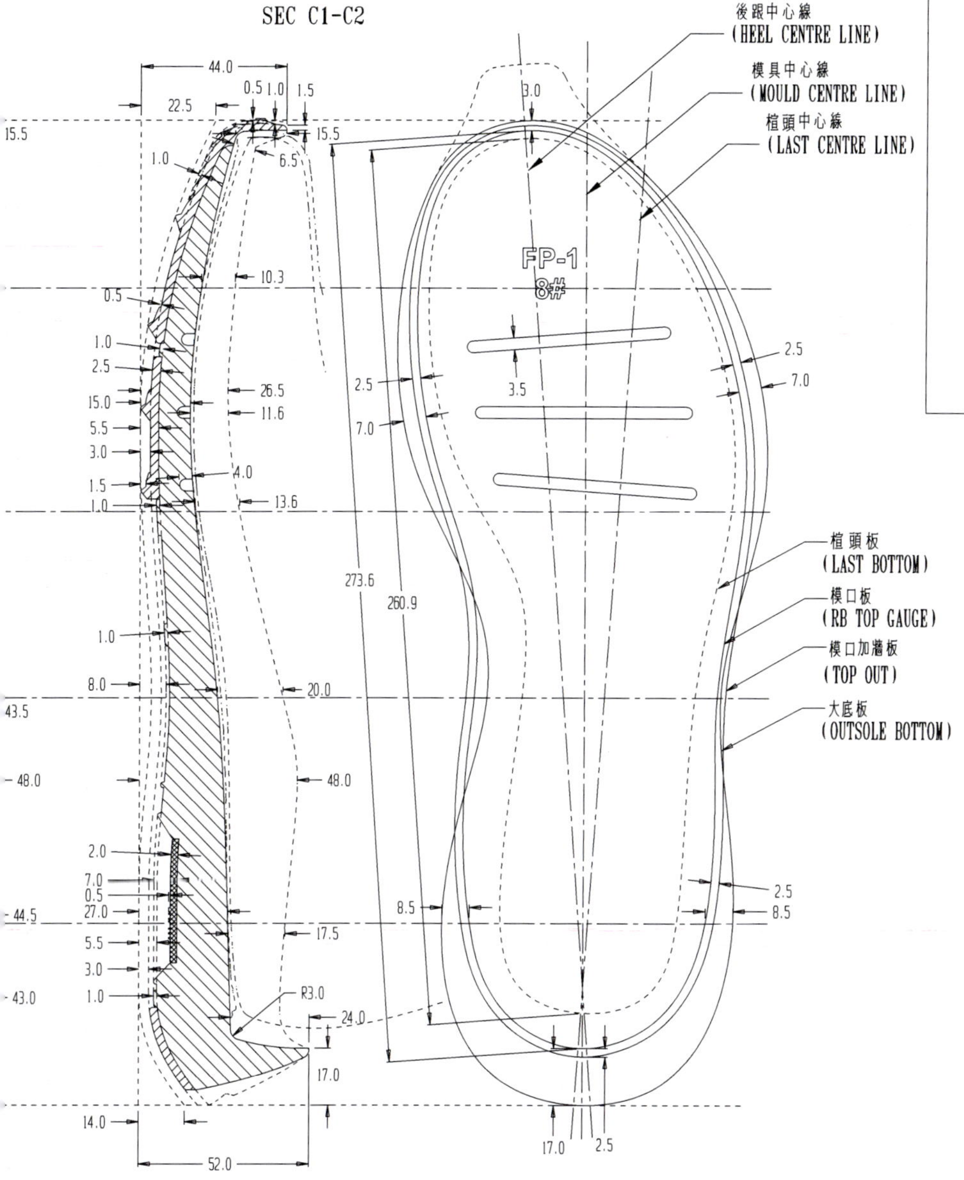

SEC A1-A2

TOP VIEW

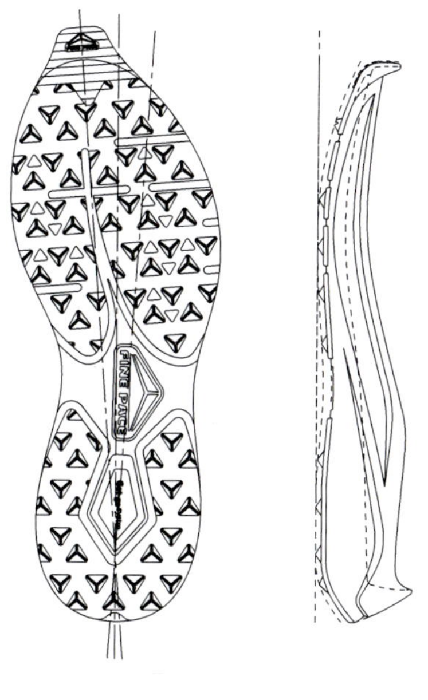

SCALE:1/2

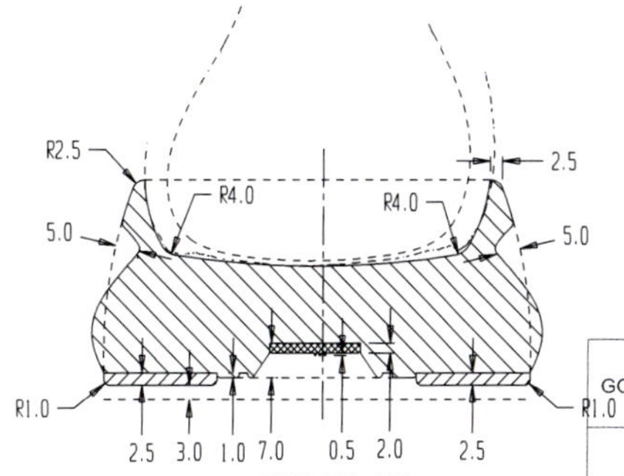

SEC E1-E2

中国源胜五金模具有限公司
YUAN SHENG MOULD CO.,LTD
TEL:0084-0938971815 0086-139 2554 4918
E-MAIL:Fish@yuanshengmold.net
越南虎胜(模具)责任有限公司
HU SHENG MOULD CO.,LTD
E-MAIL:yuanshengmold1216@yuanshengmold.net

GOMP.MATL	LIST HARDNESS	DRAWING KEY
EVA	55 c°	
RB	50 A°	

MOLD CODE 模具编号	PISA SPLT-1 LF-02	DATE 日期	2024/09/03 2025/02/15	LAST NO. 楦头号	PISA SPLT-1
DRAFTSMAN 绘图人	LH	SCALE 比例	1:1	SIZE 号码	40#
GUEST 客户	FINE PACE	UNIT 单位	MM	MATERIAL 材质	RB+EVA
审核		样品图 客戶簽名		量产图 客戶簽名	
SEASON	Summer 26	BROUGHT TO YOU BY MANUFACTURING GROUP LLC			

104

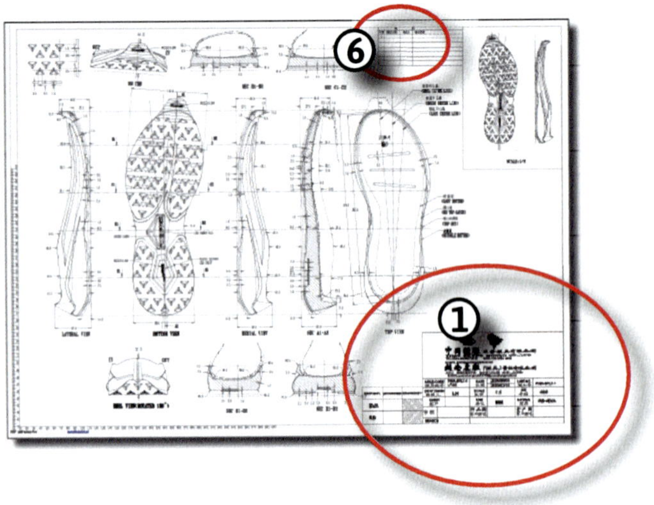

Look At The Title Block:

1. The item model name
2. The mold shop name and customer name
3. The shoe last code number
4. Drawing part size
5. The material call out
6. Revision list
7. Component hardness specs or color information

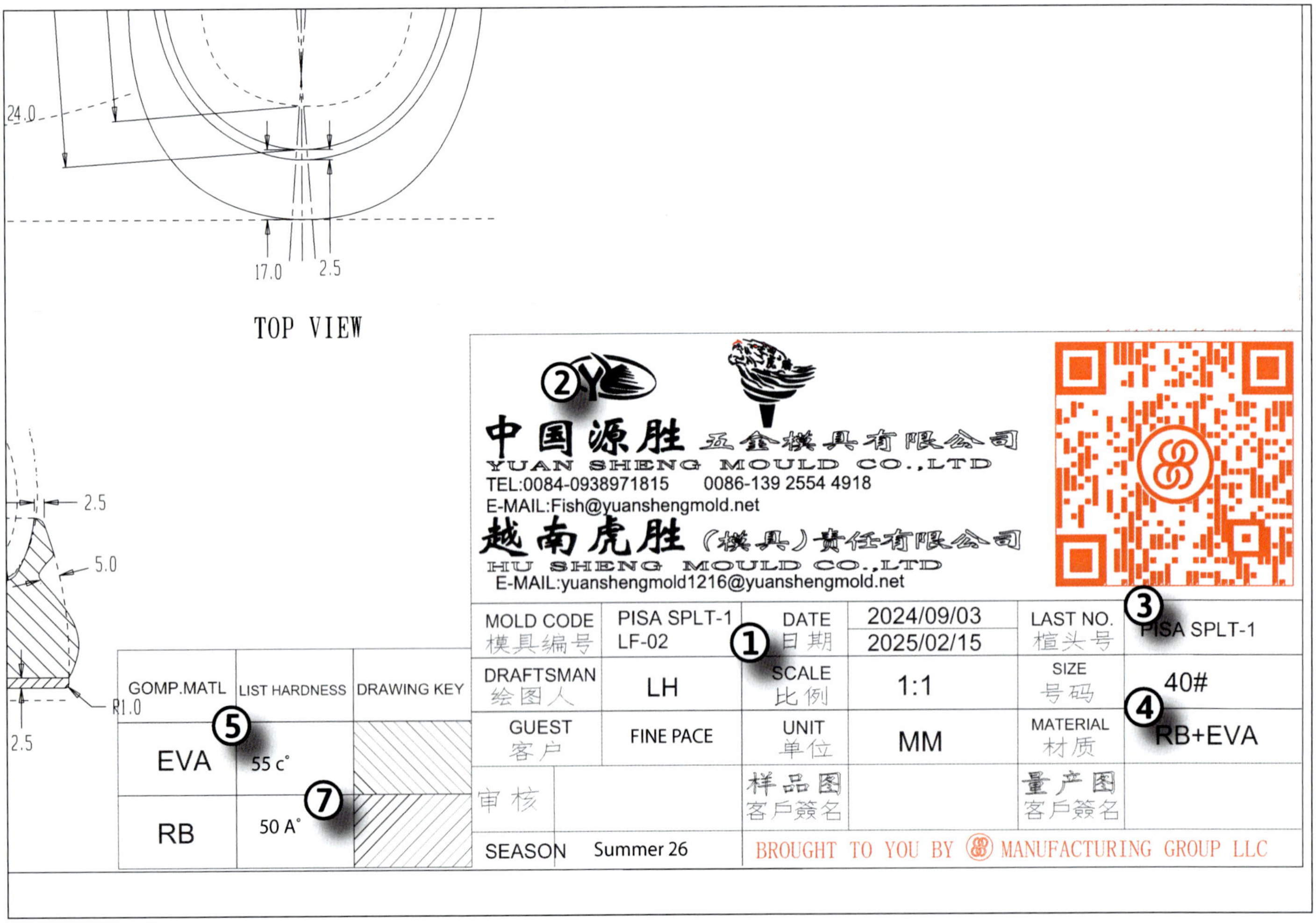

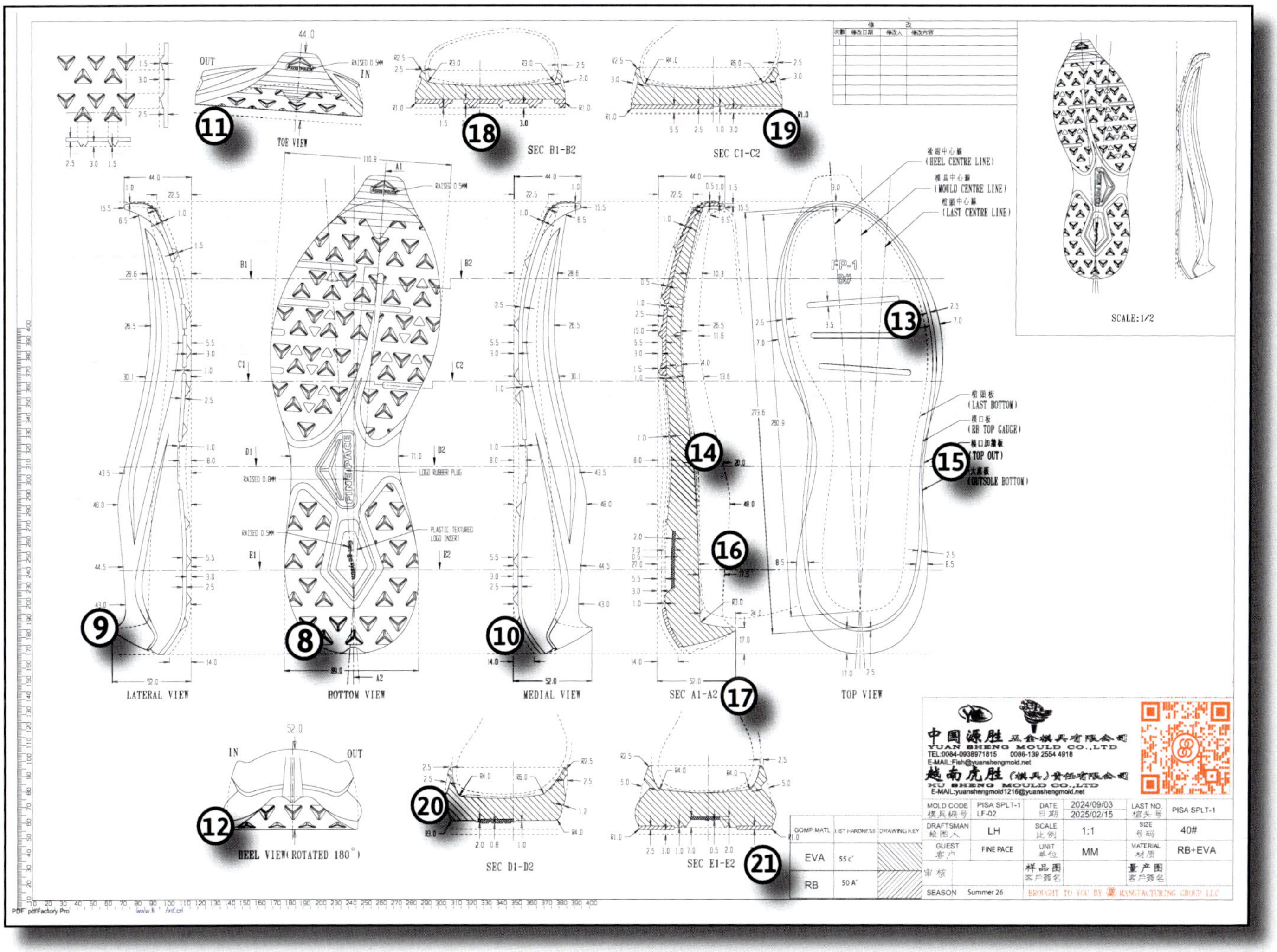

On The Drawing You Should See:

8. Bottom view tread
9. Lateral side view (outsole)
10. Medial side view (inside)
11. Toe view
12. Heel view
13. Top net - top surface (top net) view of the midsole
14. Last bottom contour line
15. Rubber top line
16. Midsole top line
17. Cross section A-A: toe to heel
18. Forefoot section B-B
19. Arch section C-C
20. Waist section D-D
21. Heel section E-E

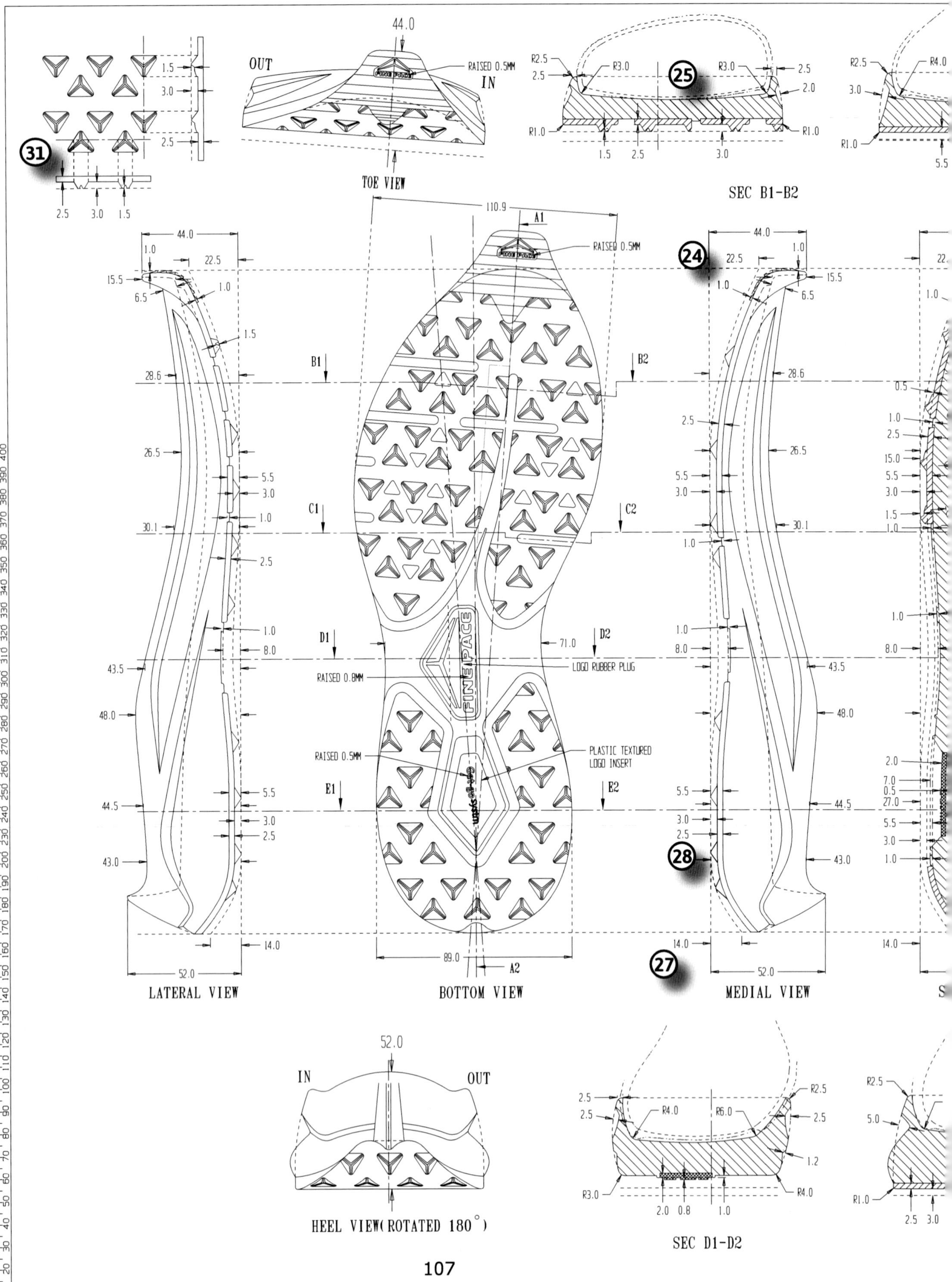
OUT
IN
RAISED 0.5MM
TOE VIEW
SEC B1-B2
RAISED 0.5MM
LOGO RUBBER PLUG
RAISED 0.8MM
RAISED 0.5MM
PLASTIC TEXTURED LOGO INSERT
FINE PACE
LATERAL VIEW
BOTTOM VIEW
MEDIAL VIEW
HEEL VIEW(ROTATED 180°)
SEC D1-D2

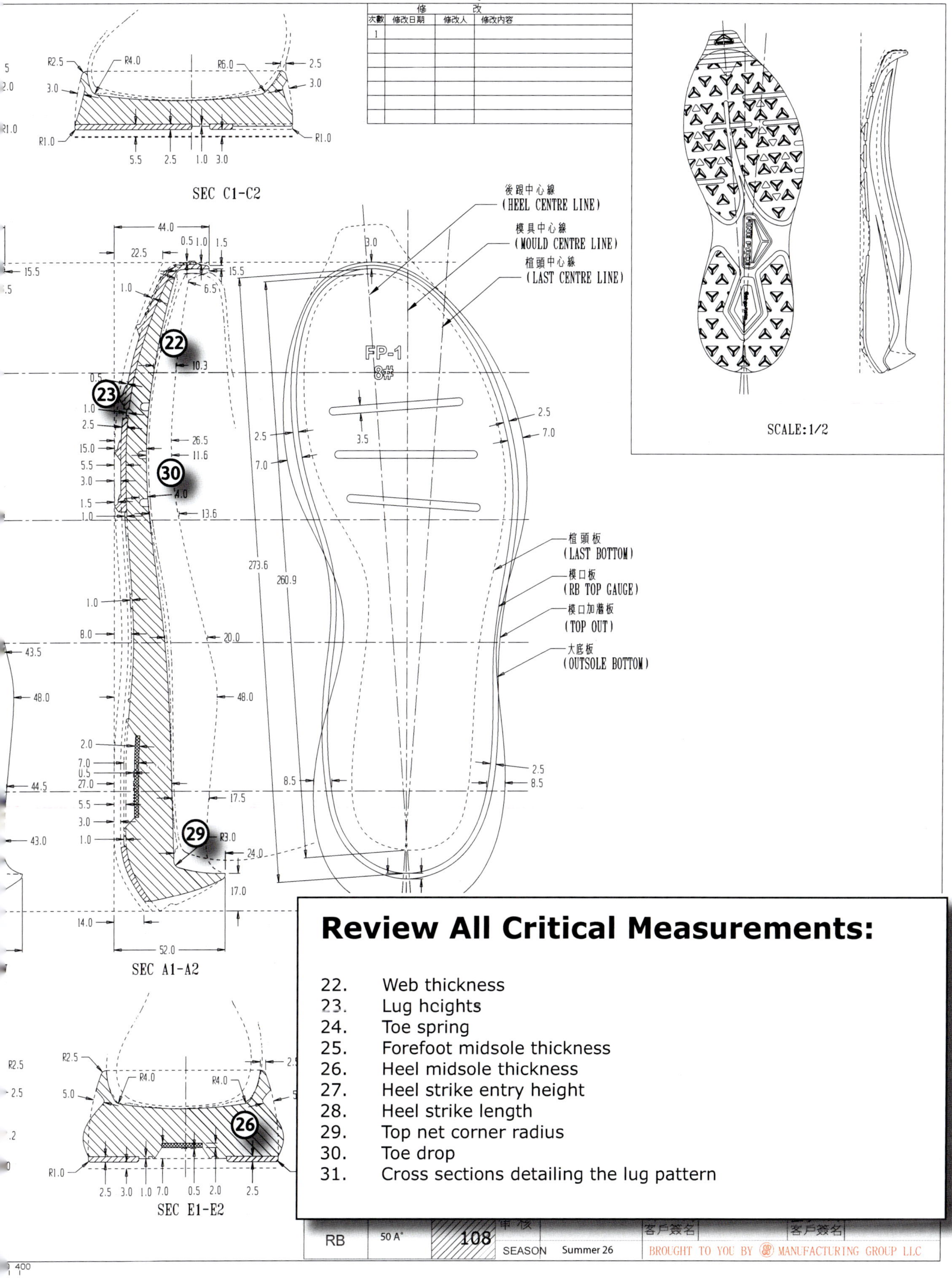

Review All Critical Measurements:

22. Web thickness
23. Lug heights
24. Toe spring
25. Forefoot midsole thickness
26. Heel midsole thickness
27. Heel strike entry height
28. Heel strike length
29. Top net corner radius
30. Toe drop
31. Cross sections detailing the lug pattern

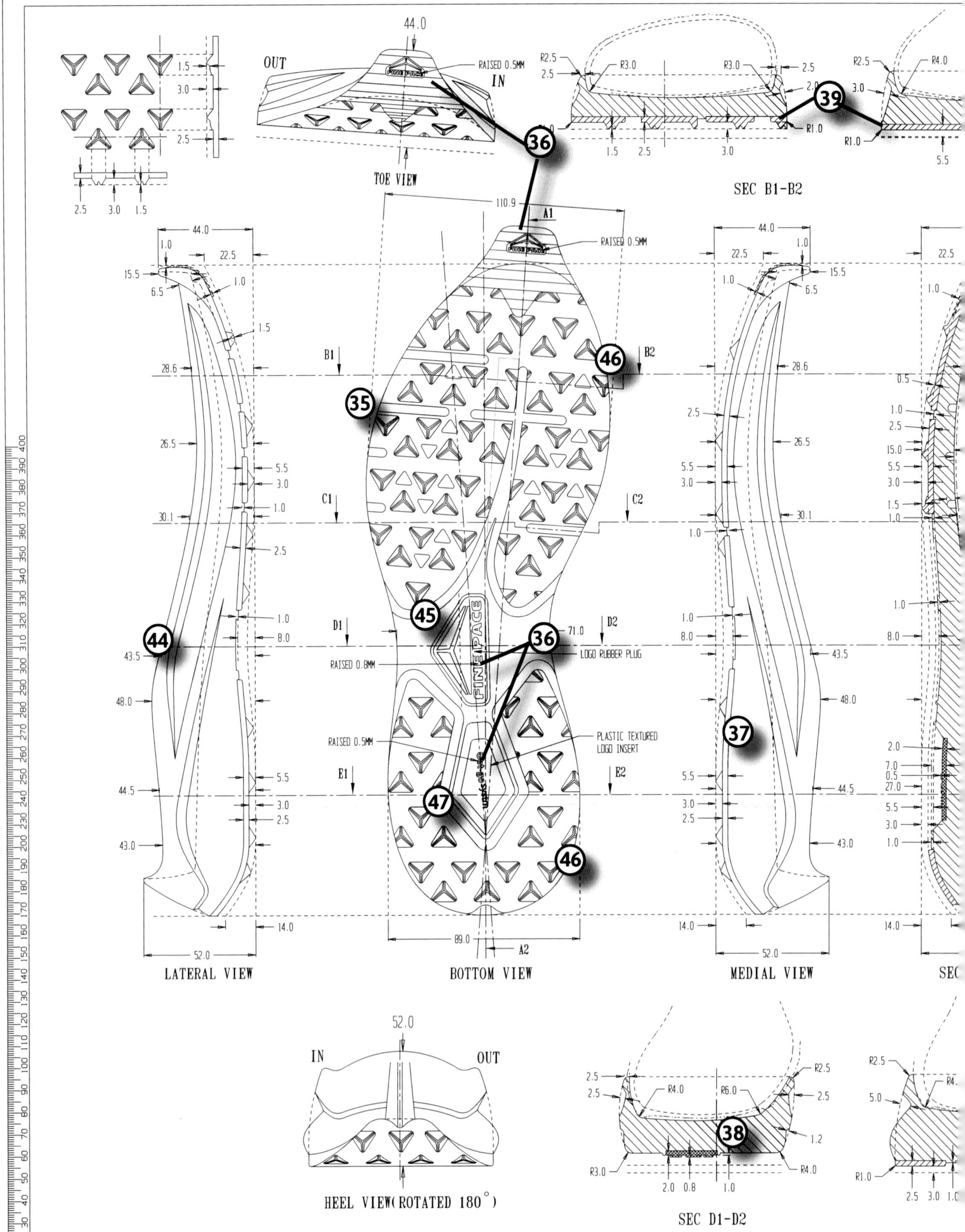
OUT
IN
RAISED 0.5MM
TOE VIEW
SEC B1-B2
R2.5
R3.0
R4.0
R1.0
110.9
A1
B1
B2
C1
C2
D1
D2
E1
E2
A2
71.0
LOGO RUBBER PLUG
RAISED 0.8MM
PLASTIC TEXTURED LOGO INSERT
89.0
LATERAL VIEW
BOTTOM VIEW
MEDIAL VIEW
HEEL VIEW(ROTATED 180°)
52.0
SEC D1-D2
R6.0
35
36
37
38
39
44
45
46
47

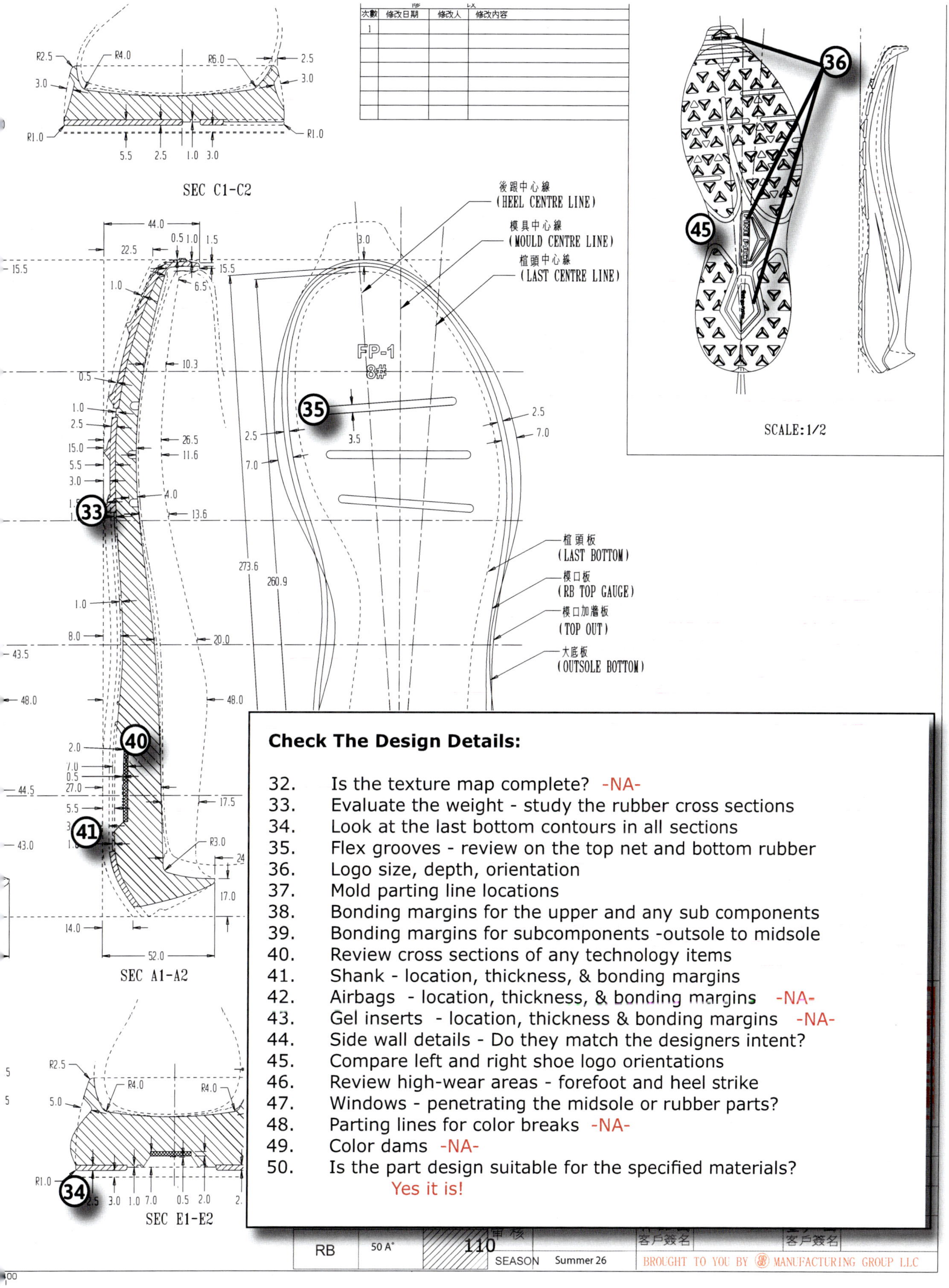

Check The Design Details:

32. Is the texture map complete? -NA-
33. Evaluate the weight - study the rubber cross sections
34. Look at the last bottom contours in all sections
35. Flex grooves - review on the top net and bottom rubber
36. Logo size, depth, orientation
37. Mold parting line locations
38. Bonding margins for the upper and any sub components
39. Bonding margins for subcomponents -outsole to midsole
40. Review cross sections of any technology items
41. Shank - location, thickness, & bonding margins
42. Airbags - location, thickness, & bonding margins -NA-
43. Gel inserts - location, thickness & bonding margins -NA-
44. Side wall details - Do they match the designers intent?
45. Compare left and right shoe logo orientations
46. Review high-wear areas - forefoot and heel strike
47. Windows - penetrating the midsole or rubber parts?
48. Parting lines for color breaks -NA-
49. Color dams -NA-
50. Is the part design suitable for the specified materials?
    Yes it is!

If you see something.....say something!

If you see something that does not look right, or you are not sure about a detail, then STOP. Ask for some help. Don't hesitate to ask the drawing technician, the factory developer, your co-workers or The Shoe Dog. Pen and ink is easy to change but once you get into metalworking it is a lot harder to make adjustments.

How much is all this scanning and drawing work?

This development process is typically broken down onto two parts.

#1 The 2D drawing work, $200
#2 The 3D drawing work, $400

If you are a huge client or repeat customer you may never see a bill for these services.

You must be wondering why these services are so inexpensive. The tooling shop knows that if they don't do the drawings, they will never get the order for the equipment. It's in the tooling shop's best interest to have a good relationship with the factory so they can get more business creating equipment for the shoe factory's customers.

As a brand, it's a good idea for you to have a relationship with the tooling factory.

Many mold makers or tooling shops will return the drawing charges if the sample mold turns into a production tooling order.

Can you get a discount if your own staff does the 2D and the 3D drawing? The answer is no. Because the shoe factory and the tooling factory are responsible to make sure that the uppers and the outsoles and the outsole sub components all fit together correctly, they will recreate any drawing that you send to them.

This is how it should be. The brand does not want to be responsible for how the parts fit together.

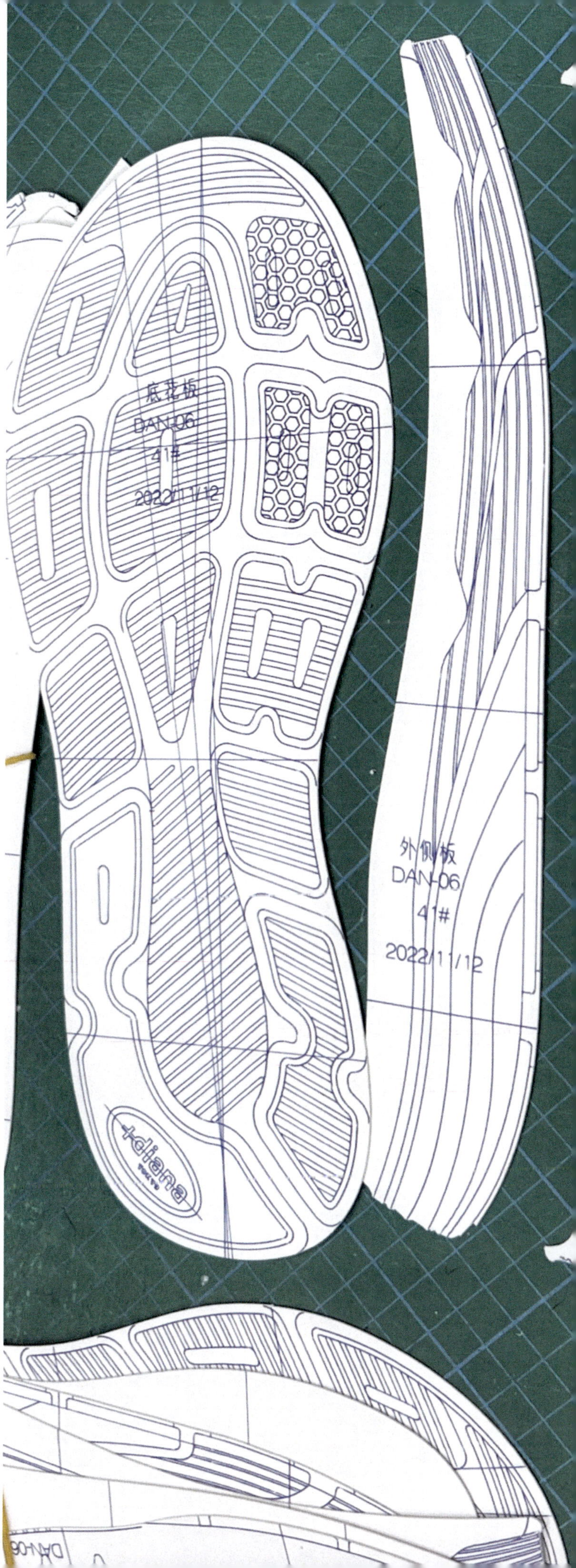

Chapter 8

Building Prototype Tools

Tooling is expensive.

Before we open a new tool we really need to review the design in 3D.

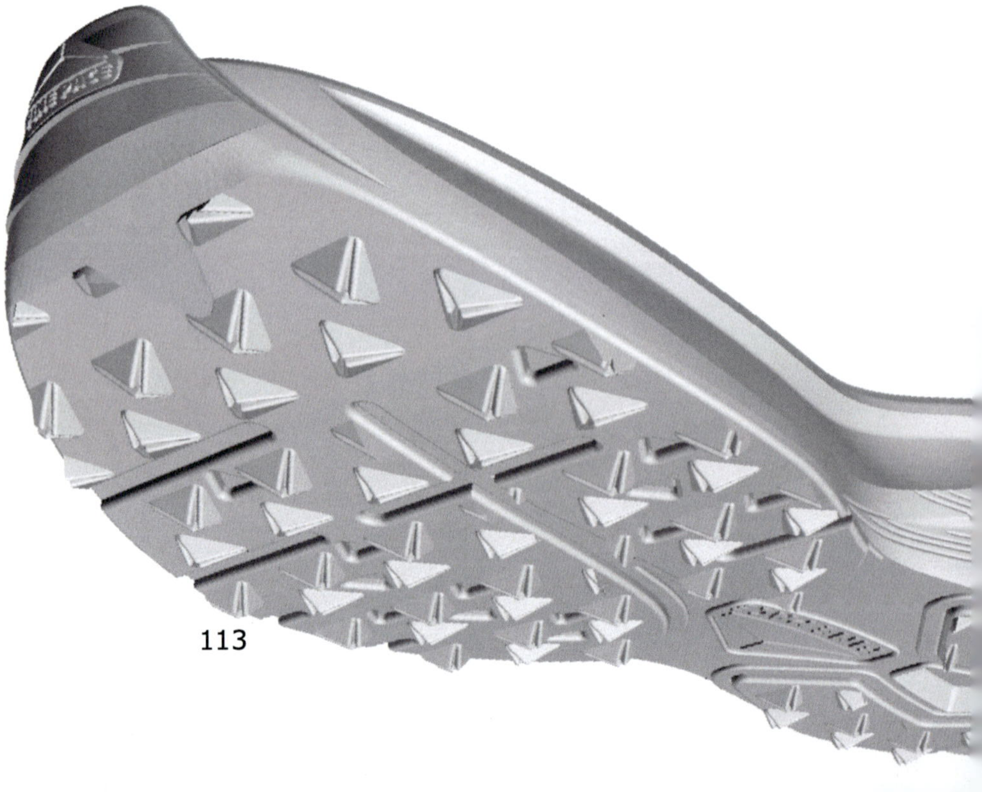

Prototype tooling

After you study the 2D blueprint and make the revisions, it's time to make the 3D CAD model. When reviewing the 3D CAD model, it's important to shift your focus to the styling intent of the shoe designer. The 2D blueprint is fantastic for looking at cross-sections, engineering complications, and bonding margin. All of these are technical details that will make your shoe design work in a functional piece of footwear.

The 3D drawing is now a great opportunity to look at every curve, contour and style line. Midsole contours are difficult to render in 2D so now is your chance to really make sure the designer's intent is met. Of course, check all your logos.

Again, just like the 2D blueprint, if you see an issue, don't hesitate to make corrections. CAD files are easy to manipulate. Remember, modifications to the 3D file may require some adjustments to the 2D blueprint.

Once you've confirmed the 3D CAD model, it's time to make the wood cut. Wood cut is an old-fashioned term. Today we use filled resin blocks or a 3D printer to create the hard model of the outsole.

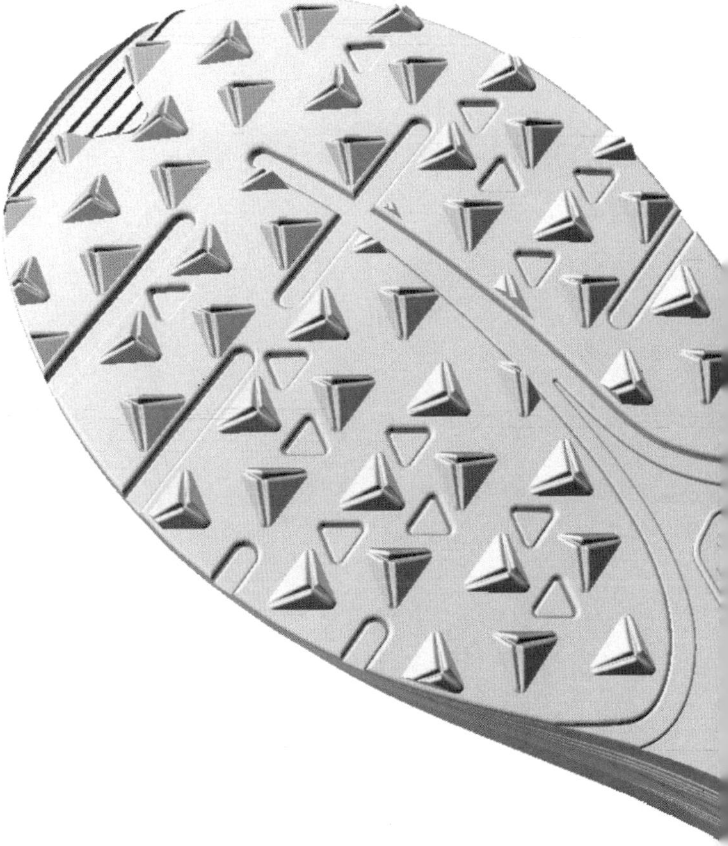

CNC cutting the "wooden" mold

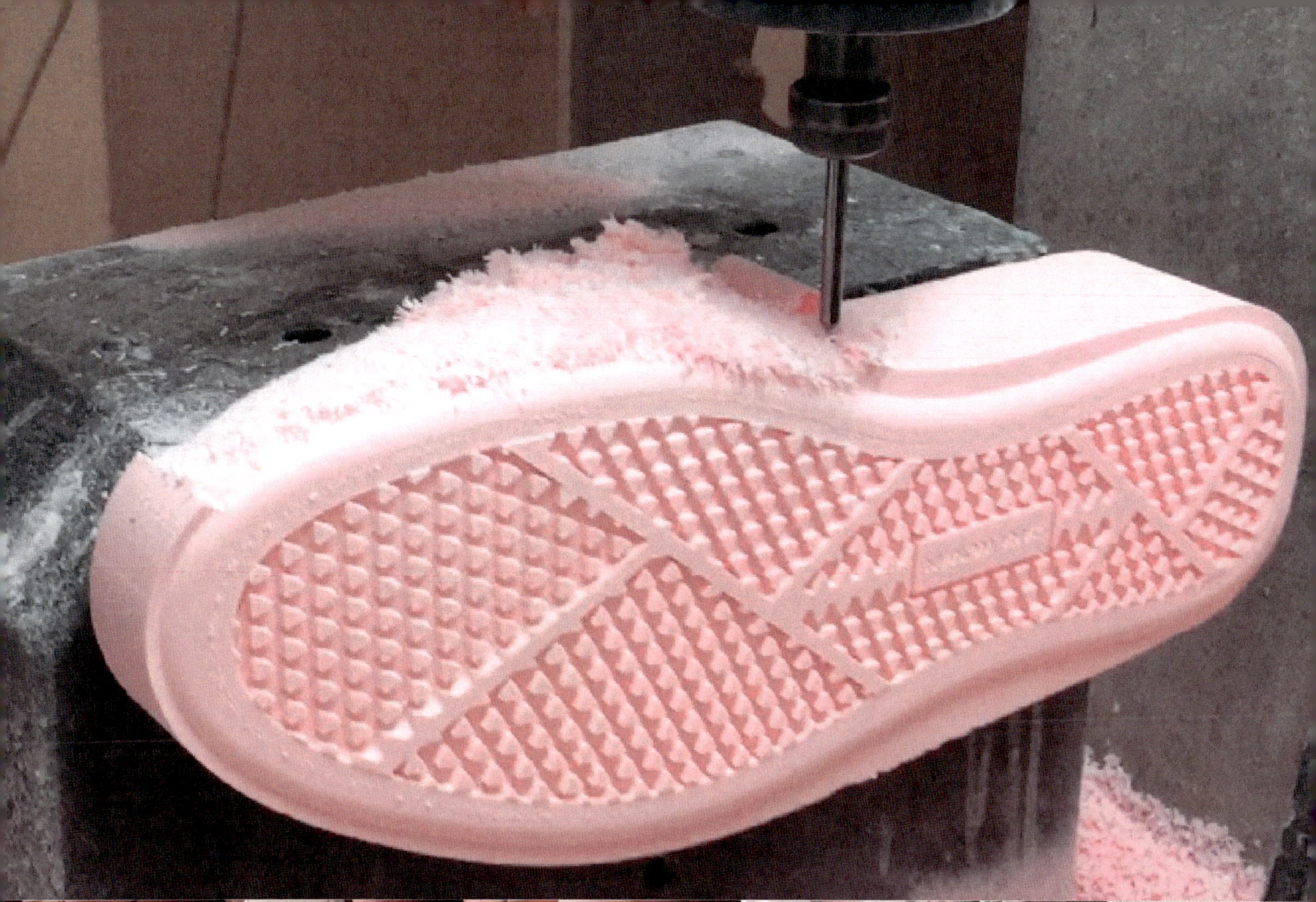

Clarks

SLA - stereolithography
3D printing

SHENG NO

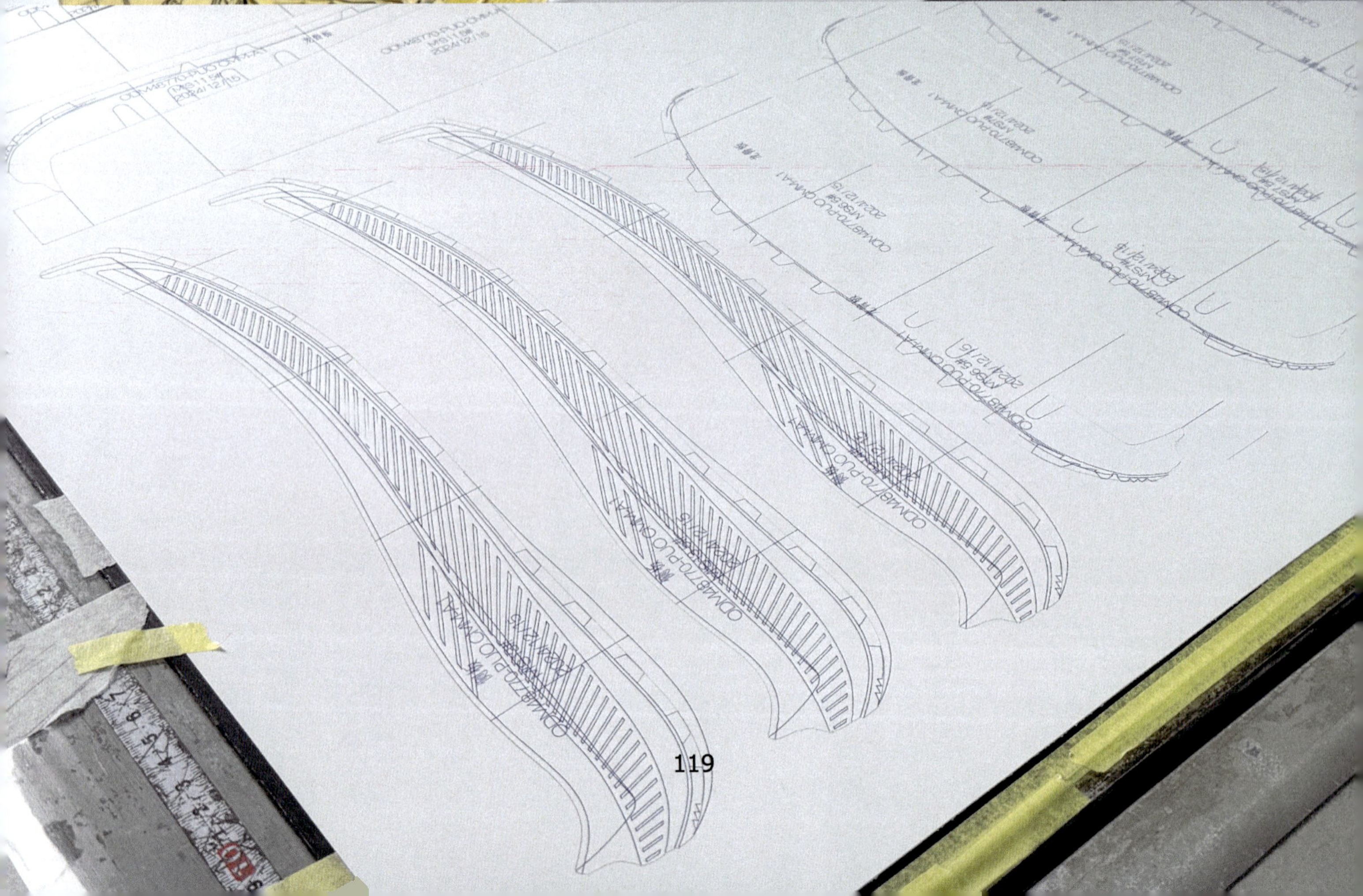

Review the wood cut model

Reviewing the wood cut is an important part of the process to once again confirm the final 3D hard model is exactly what you had in mind for the design. If you see something that's a problem, it's very important to correct it even if you have to make a new wooden model.

Open up your 3D file, get out your 2D blueprint, get the wood cut and a ruler and check everything. This is the last stop before you start cutting metal.

Can you skip this step to save time? The answer is no. The wood cut is actually used during the mold manufacturing process for metal molds made using the casting process.

During product development, you'll be making just one size, but once you confirm the design in order production, there will be a wood cut made for every size of mold.

Mold technicians in the factory will review every size to prepare the equipment for casting. It's not uncommon to see a technician filing, sanding, or using a little putty to repair the wood cut in preparation for the casting operations.

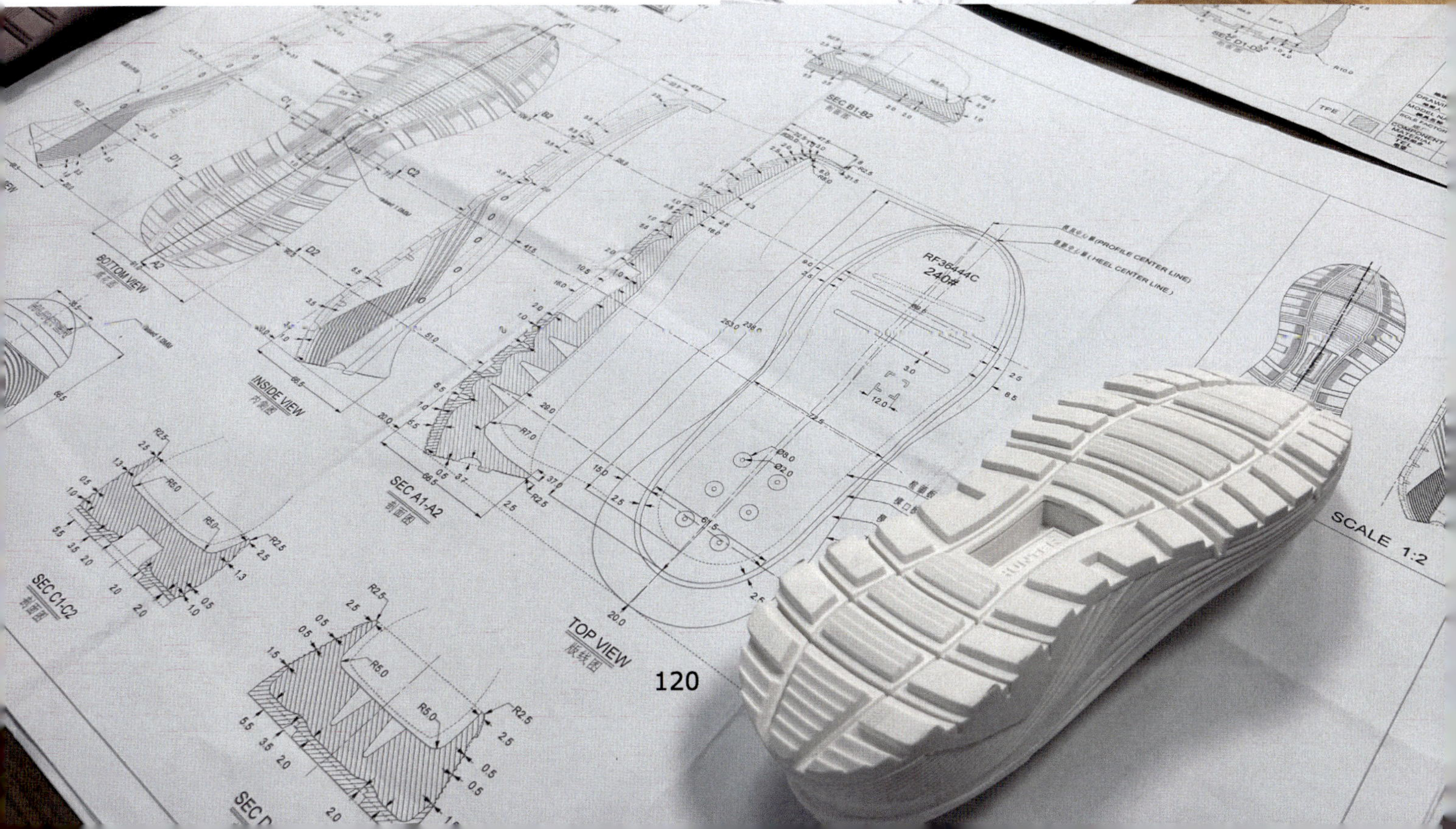

The technician must check every size

This assembly is ready for the metal casting factory

Mold Drawings Versus Part Drawings

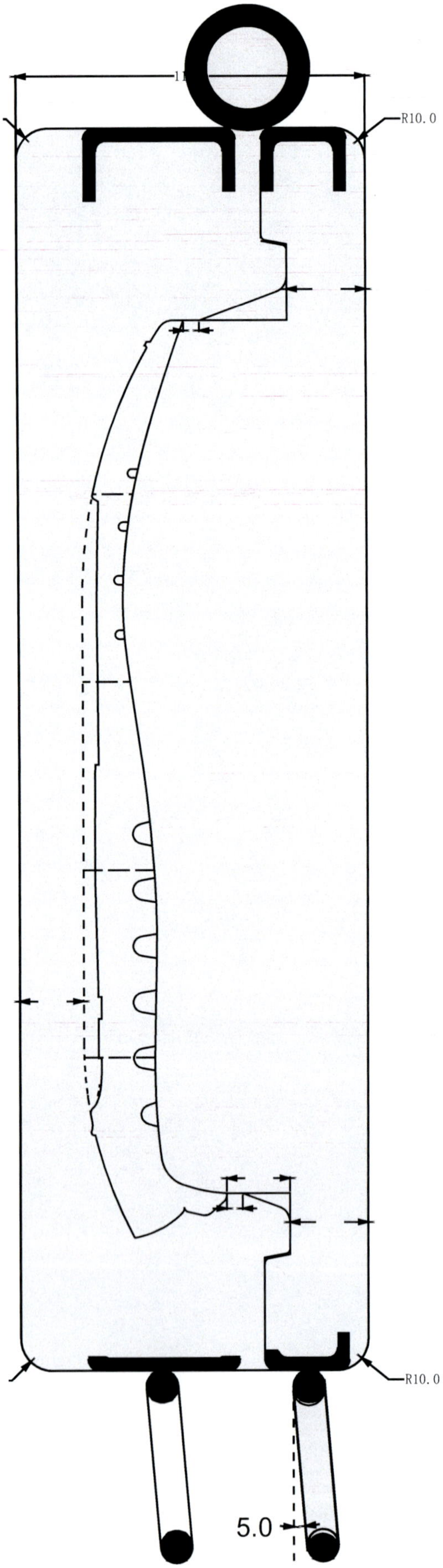

As a professional footwear designer or footwear developer, you may never actually see the mold drawing. Your efforts will be 100% focused on the part drawing. For conventional footwear tooling, it's just not important.

However, if you are working on a new technology that requires some new or special technique, you definitely will be looking at the mold drawings.

It's too easy for a designer to create something that cannot be manufactured, so, as a developer, it's important for you to understand exactly how these molds operate and what machinery is required with the mold to produce the parts.

Now, more than ever, modern automated manufacturing will have strict requirements for how the tooling adapts to a particular molding machine.

The drawing for a cast aluminum EVA pressing mold.

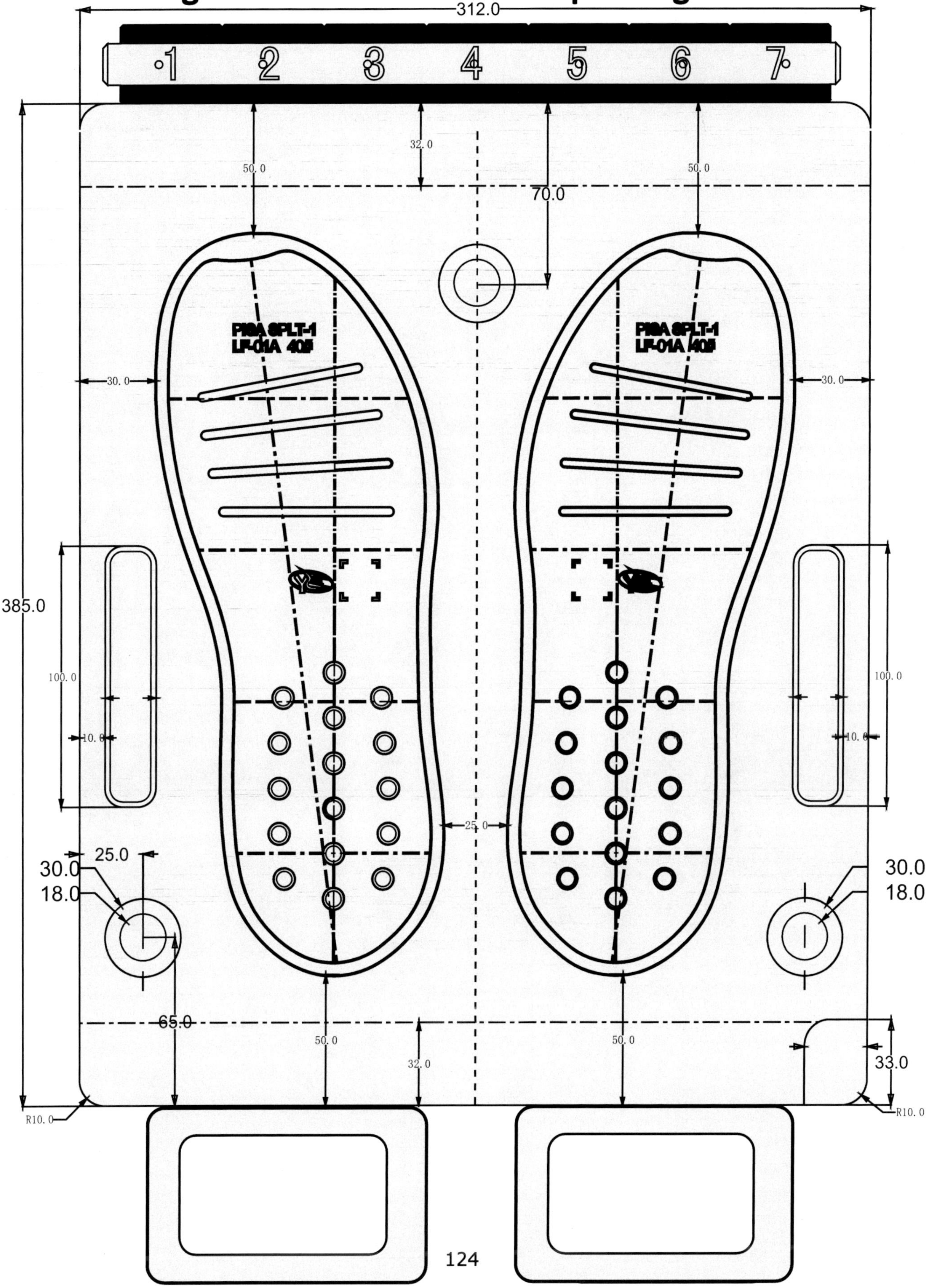

The drawing for a cast steel rubber pressing mold.

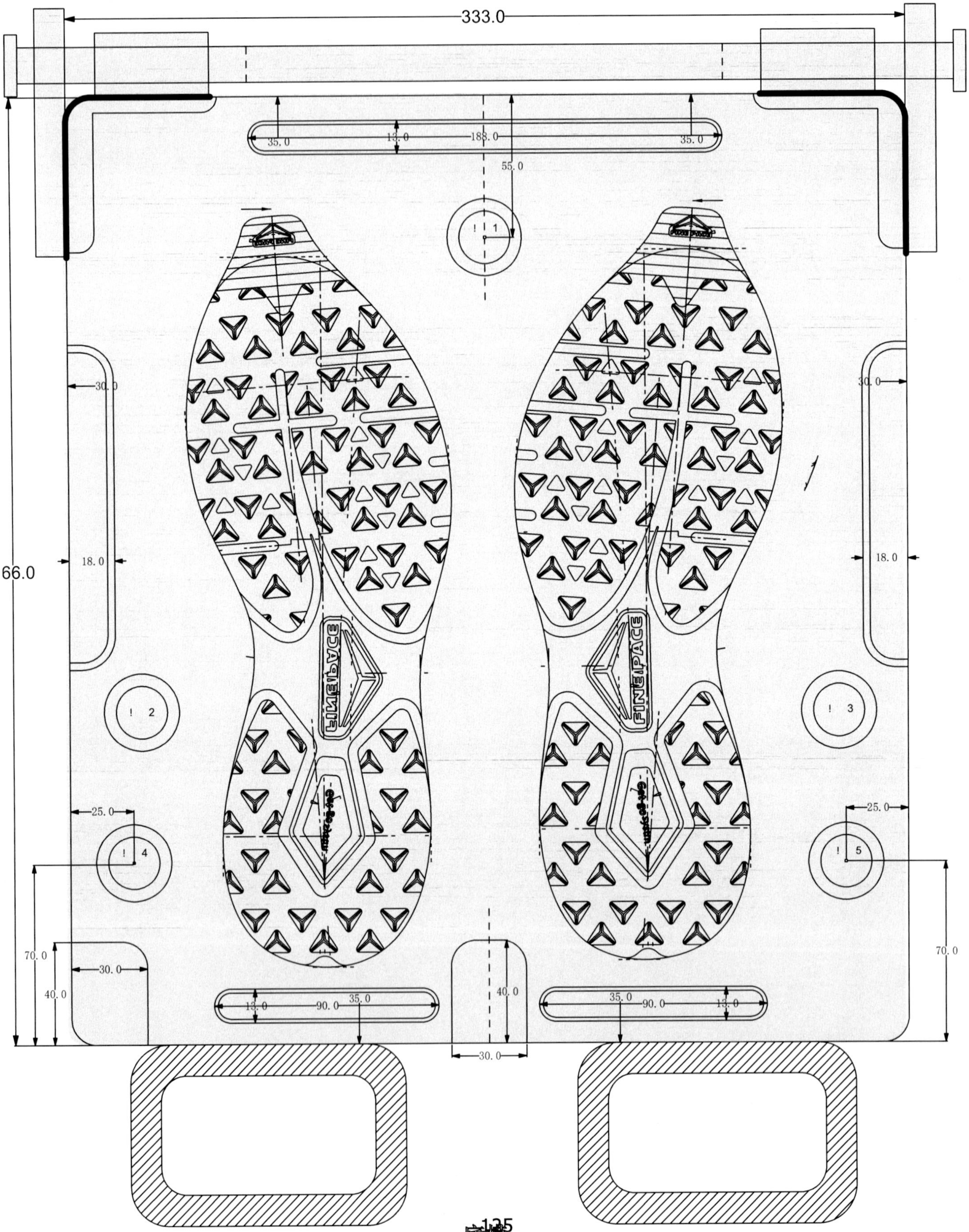

底模

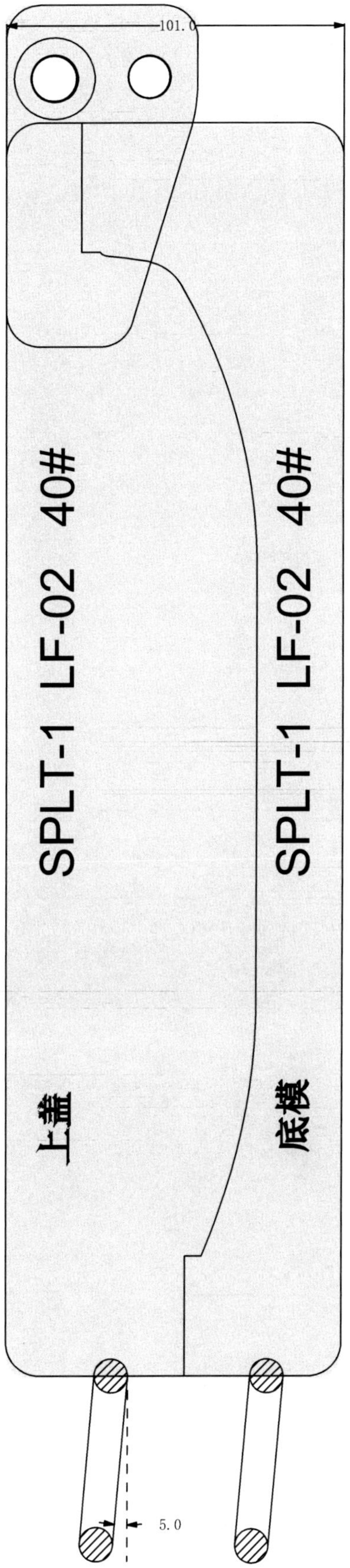

MEDIAL VIEW
内侧图

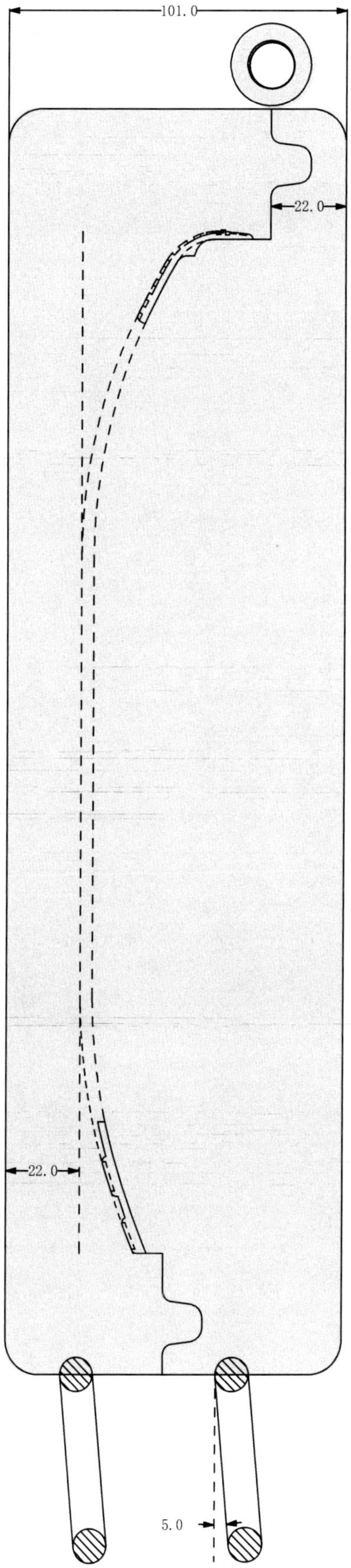

SECTION A-A'
剖面图

Pre-heating
the mold
SHIMPO

An electric arc furnace heats the steel

Pouring molten steel, 2500˚F

Aluminum is cast at 900˚F

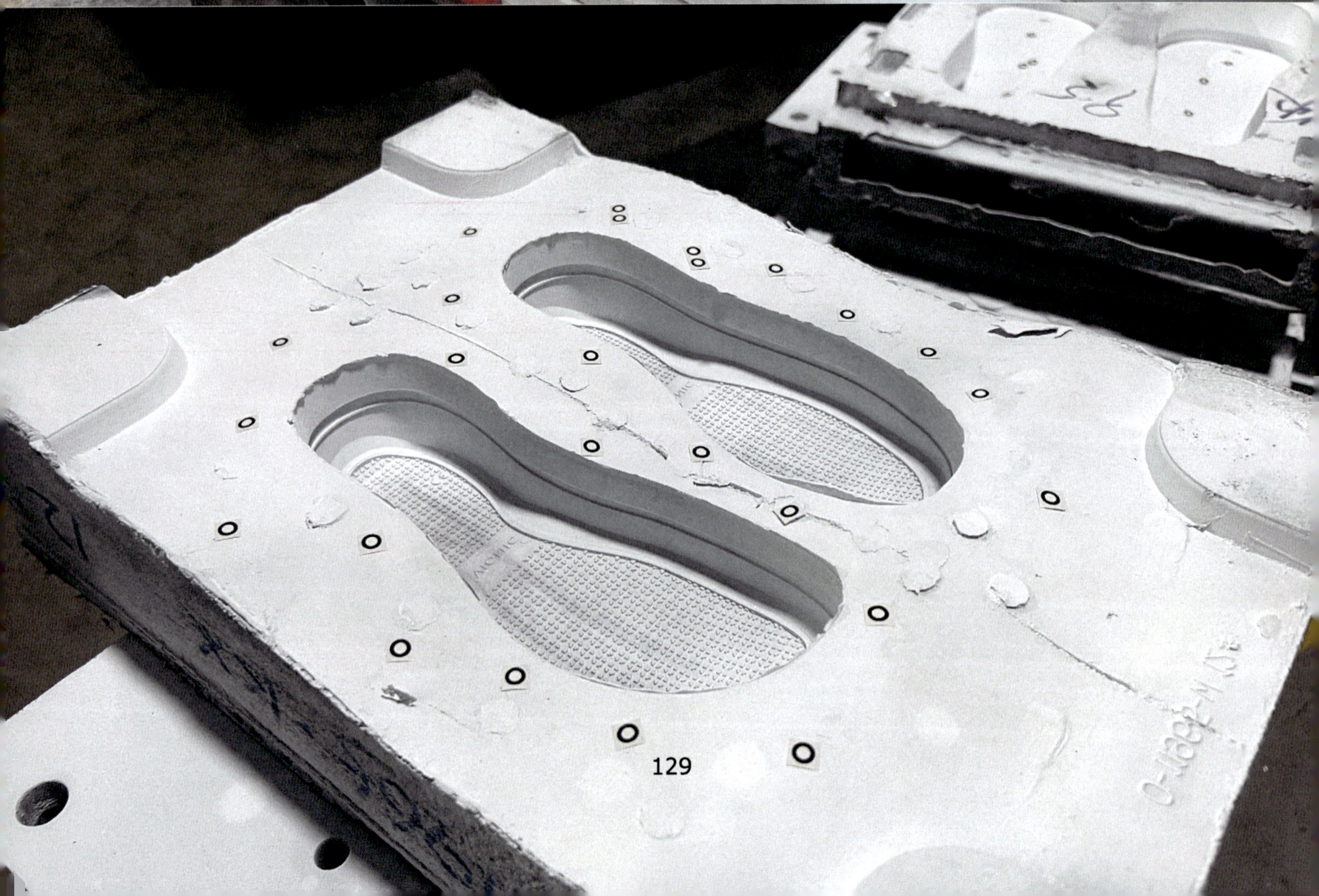

Raw castings are scanned

Process testing the cast of metal molds
The metal casting takes place at another subcontractor outside of the mold making factory. Metal casting work is dangerous and difficult work and is done by a specialized supplier.

Once the cast of aluminum and metal blocks arrive back at the mold shop, it's time for processing to turn them into real molds. First, the raw casting blocks are scanned. This ensures, the mold cavities are correct and the block is suitable for the final machining operations.

The CNC machining operations guarantees that both halves fit together correctly. This ensures the mold can close tightly to make smooth, even, and thin parting lines. This also ensures the mold will fit to the molding machines correctly.

In the case of pressing molds, hinges and handles are added. In the case of injection molds, the mounting points for the injection machines need to be created.

Steel hinge parts for compression molds

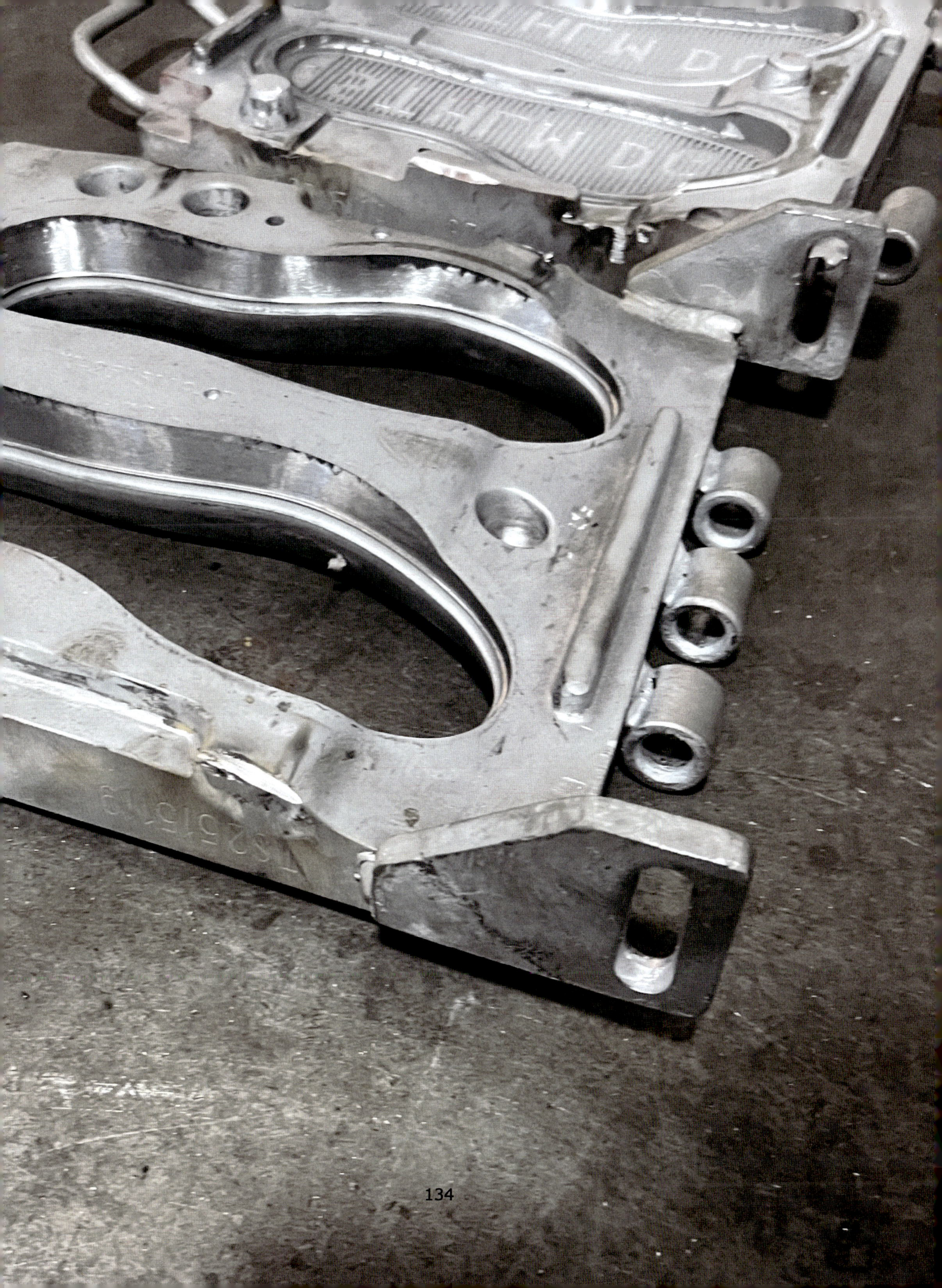

Mold clean up after test pressing.

Test pressing parts

EVA injection molds

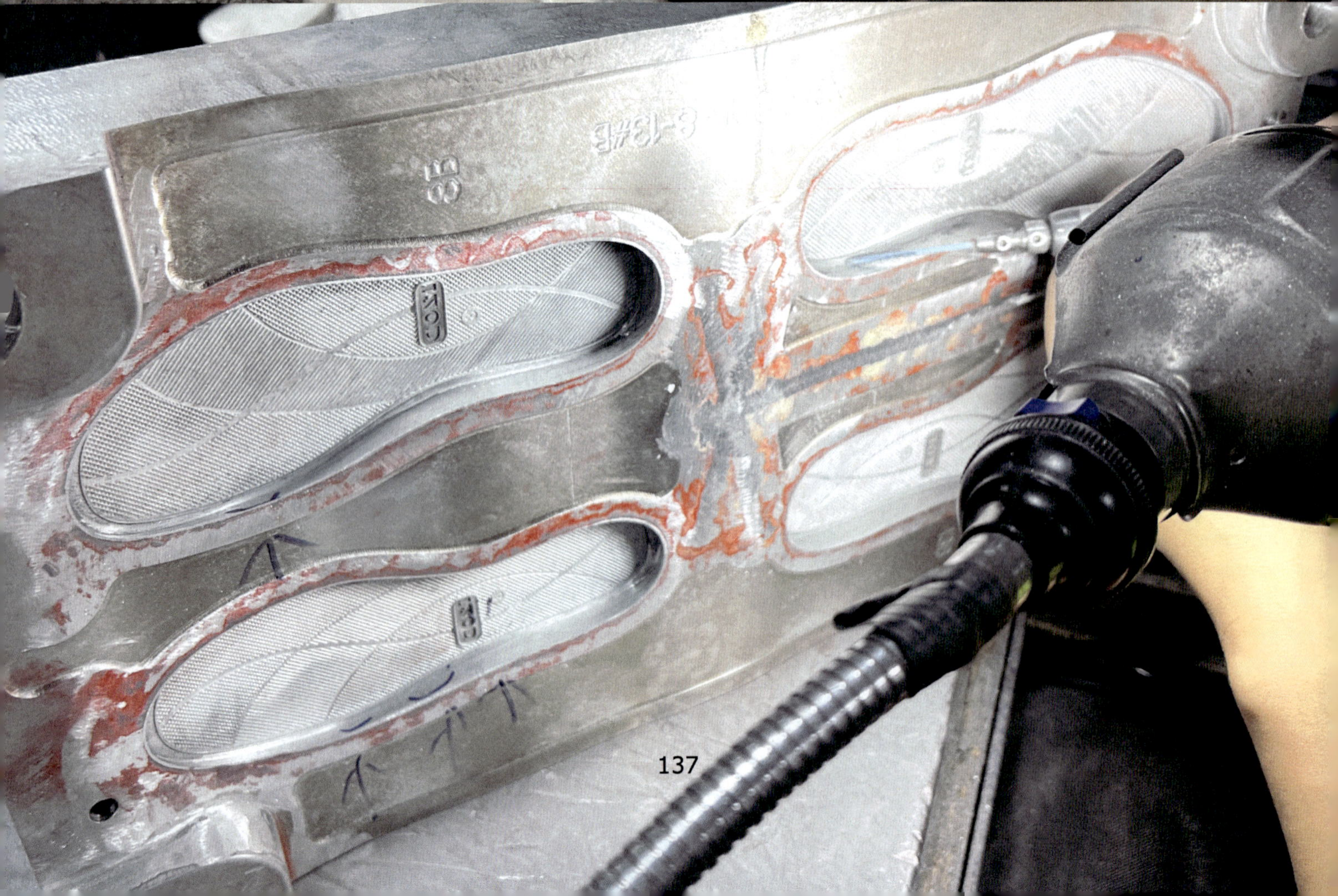

Mold Finishing

There is still lots of hard work to be done finishing the molds. After the CNC cut wood block, the casting, and the CNC cutting of the mold blocks, there is work to be done inside the mold.

In larger mold shops, you will see small banks of rubber, pressing, or EVA injection equipment. You will see a two station or a three station press. This is so the factory technicians can test the molds without having to send them out.

It's important to look at every part from every size of every mold. This ensures there are no functional defects with the mold or cosmetic problems with the parts.

Inside the mold shop, you will see lots of this red molders die. This red material is painted on the mold halves, then the mold is assembled. The red tells the technician if there are any high spots. It's important to make sure the parting lines close evenly and tightly.

We often see and hear the technicians pounding on a closed mold with a soft metal hammer. This hammering is what transfers the die so they can see exactly how the mold is closing. It is very noisy with hammering, grinding, and drilling.

contaGRIP

WS7.5 138.5

Complicated
parting lines require
precise CNC cutting

Mold Textures Air Vents & Features

When the mold is almost complete, the rubber factory will do a test pressing. They're testing how the parts fit together before they add the finishing touches.

Some features and details are added after the test pressing.

Chemical texture

The Test Press

When the mold is almost complete, the tooling factory will make a test press. The test press ensures the parts fit together correctly. They may send test press parts to the shoe factory to make sure that it fits the upper correctly. As the customer, you may not see the test press. If your company has staff in Asia, they will review the test press parts.

If the test press parts are good, the mold will move on to the texture application. Mold texture operations may be done inside the mold shop or sent to a specialized texture application subcontractor.

Mold texture is applied by several methods. Some larger design elements are created in the metal casting while fine elements have to be applied afterwards. Texture elements are created using an acid etch process. Acid resist is applied to the mold as a liquid painted on, or by texture release paper.

A worker will apply the chemical resist with an applicator and a paintbrush.

Micro textures can be applied with release paper. The design is printed on the paper, the paper is furnished into the surface of the mold. Once the surface is peeled away the acid resist remains. When the tool is exposed to acid, the elements painted with the chemical resist will become high spots in the mold.

These high spots in the mold will become grooves in the finished parts.

High-quality molds will have carefully applied texture or may not have any texture. Low quality molds will sometimes have heavy or aggressive textures to cover rough work.

Yes, you can have a custom texture paper printed with your company logo.

Mold Coating

After the texturing operations are complete, the mold surface is protected with a Teflon coating. The Teflon is sprayed and then cured in an oven. The coating protects the molds from rust and helps ensure parts don't stick inside the mold during production.

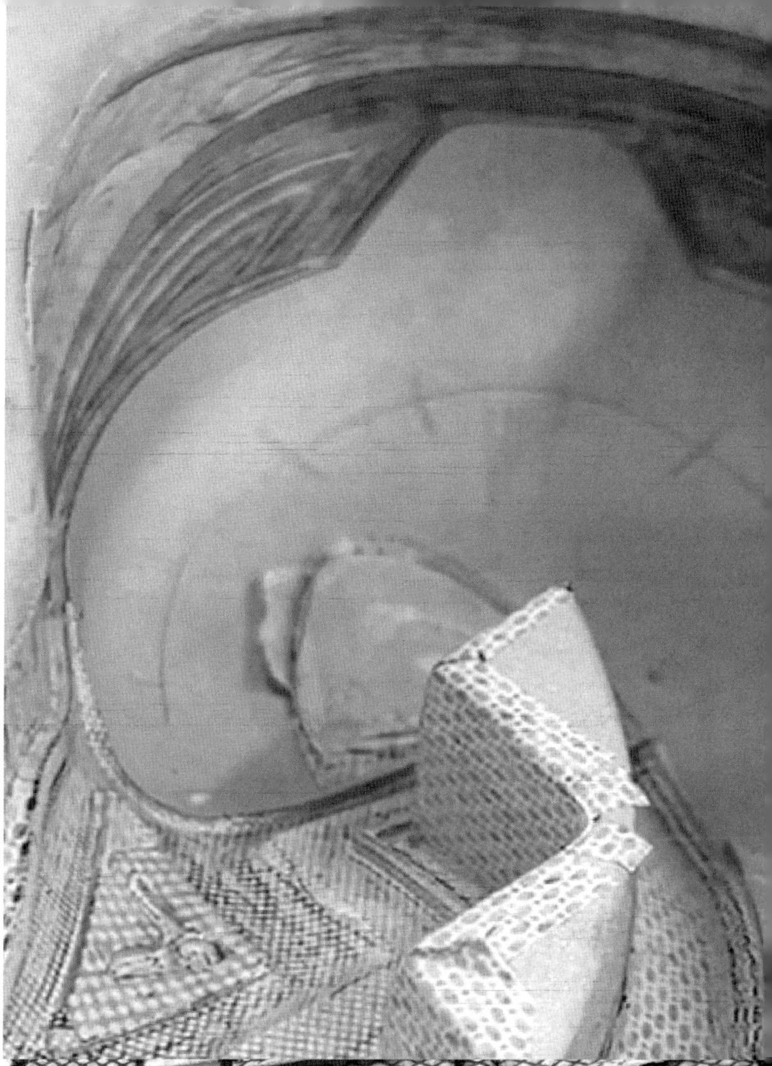

Teflon
coatings

EVA mold:
Drilling air vents

Air Vents

You will see small dimples in many EVA parts. These are the result of air vents. As a designer, I hate to see these "defects" built into my parts but, to ensure proper molding, they may be required.

These vents allow air trapped in the mold to escape. You will not see these on rubber parts or injection parts. During those molding processes the air has time to escape as the material flows into the mold cavities.

You will see in the EVA forming operations that the EVA preform is loaded into the mold squeezed or compressed into shape. As the EVA heats, it expands and remelts to fill the mold. Any trapped air will become holes or bubbles.

Here you can see the electrified wire is used to burn a fine hole through the steel. Below you can see the finished vents, to the right you can see the molded parts with dimples left behind as the EVA filled the vents during molding.

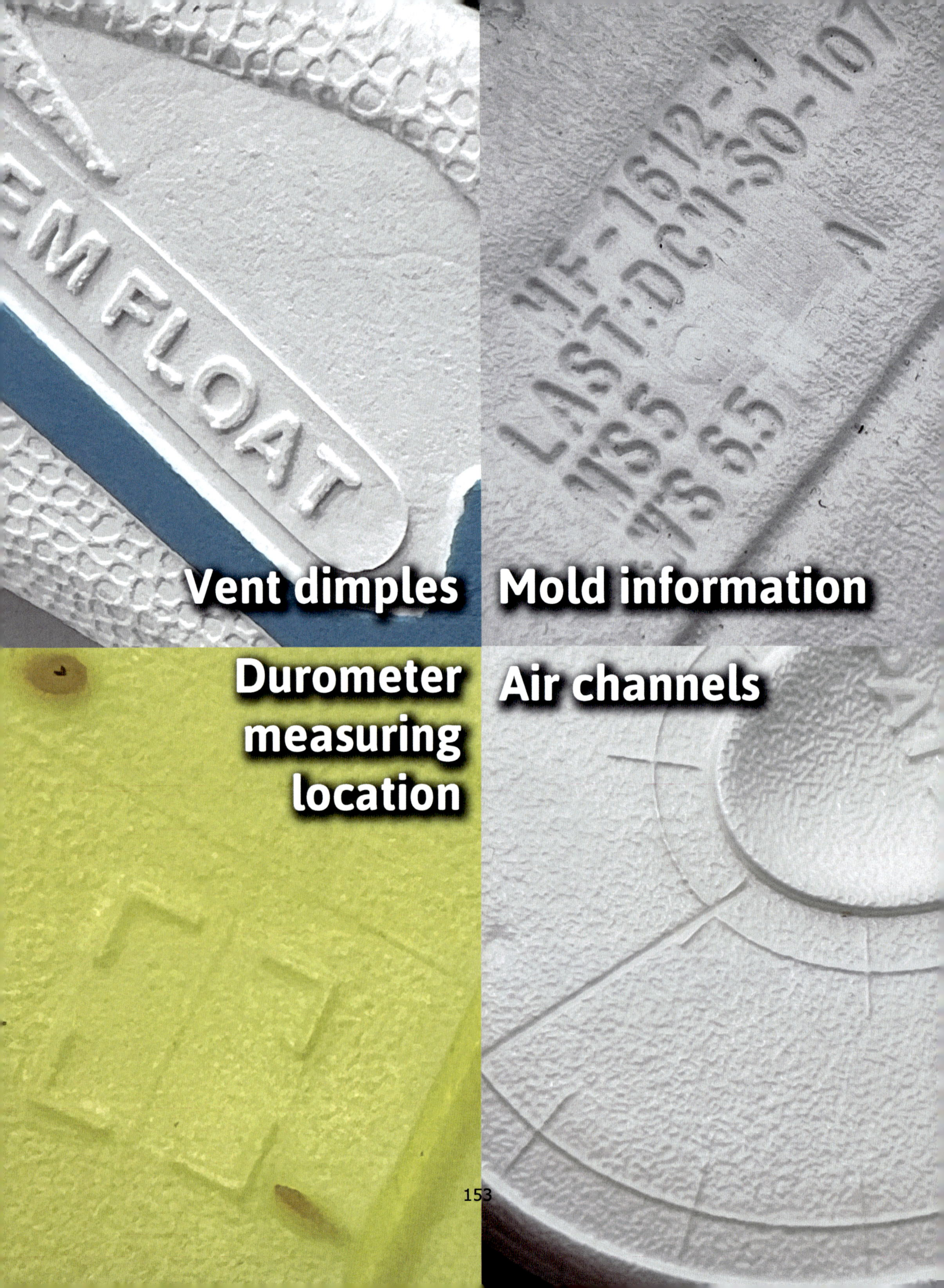
Vent dimples
Mold information
Durometer measuring location
Air channels

Air bubbles
Egg crate ribbing
Untrimmed sprue
Air vents
E-TPU "Boost"

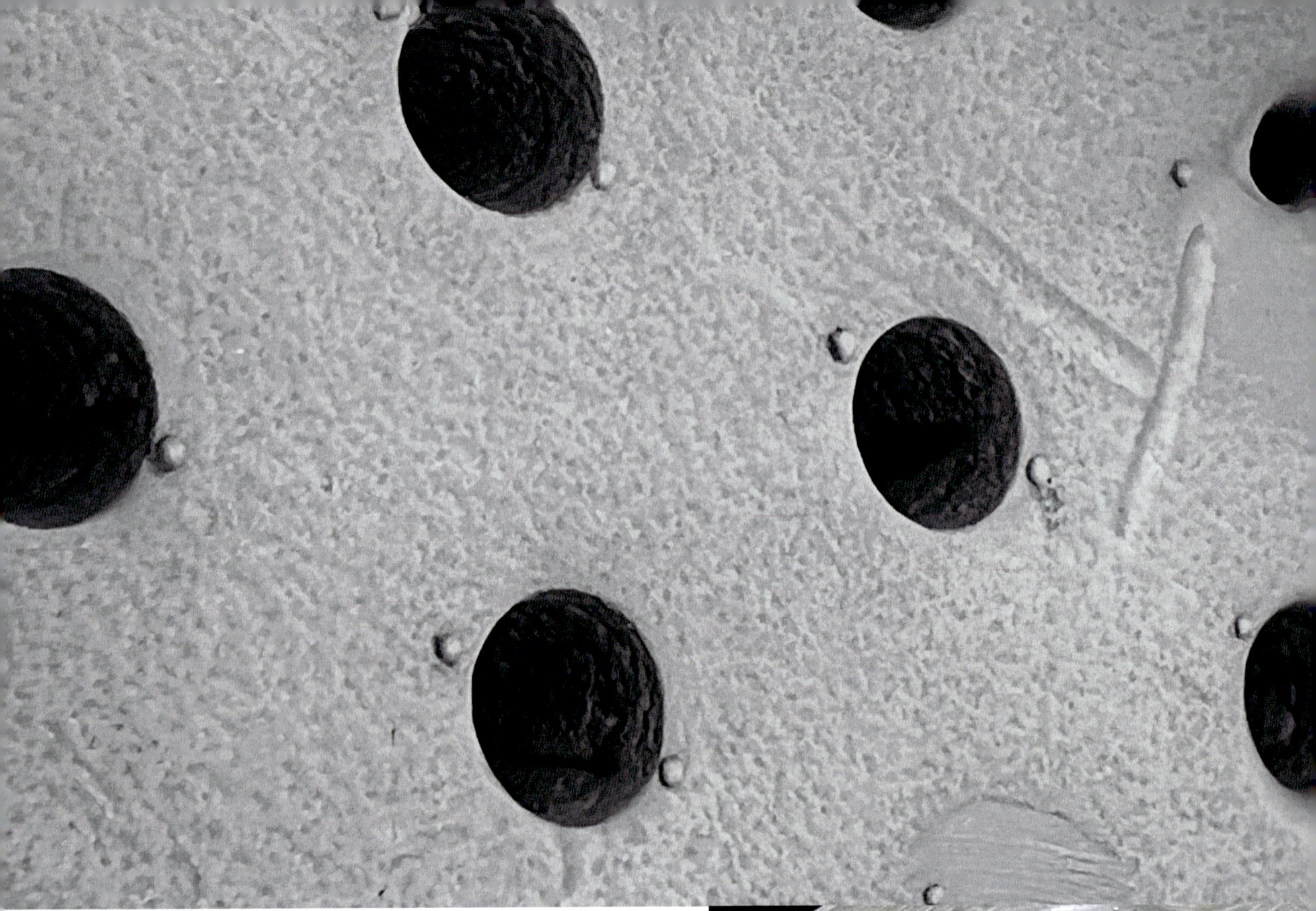

Why do EVA parts have all these holes in the heel? These holes in the midsole are created by the cone shaped projections in the top of the EVA mold.

In a PU mold, these reduce material, lighten the part, and make the heel softer. In an EVA mold, these features help heat penetrate the foam during the molding process. When EVA is molded, the material is reheated to help expand the foam to fill the mold.

These projections ensure the EVA is evenly heated so it can have even expansion and even cooling to make better parts.

You can tell this is an EVA compression mold due to the air venting lines.

Why do some molds have what looks like screw heads in the surface? You will find these screw heads in the top surface of EVA injection molds.

During the EVA expansion process, the mold opens quickly, allowing the part to expand. The undercuts on the screw head yank the EVA part out of the mold so that it can expand freely without interference from the top cavity of the mold.

So yes, they are just screws.

You can tell this is an EVA injection mold by the wide flat parting plane and the large EVA injection port between the screws.

Chapter 9

Outsole Rubber & EVA Compression Molds

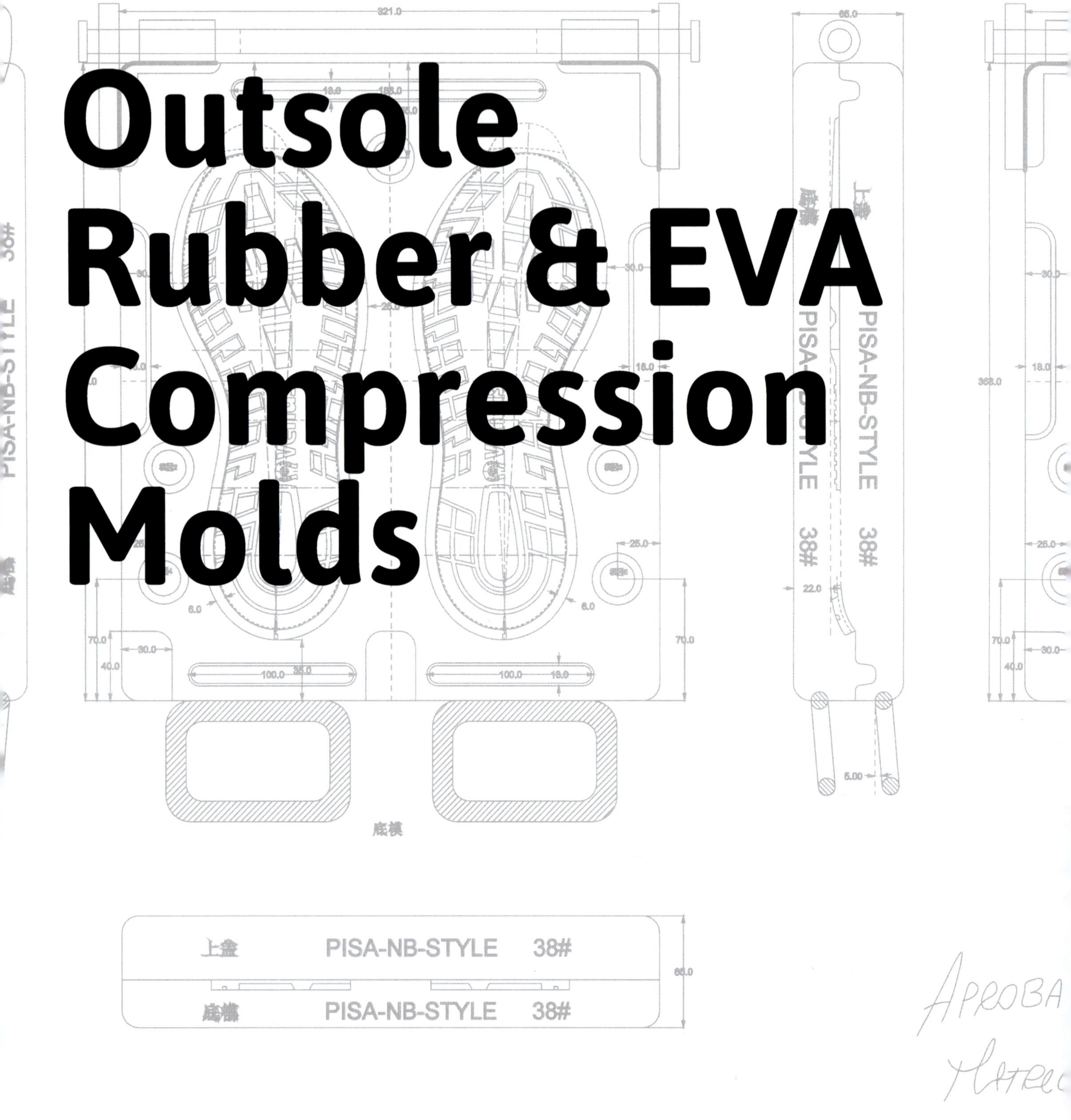

A press bank for rubber and EVA processing

Rubber Pressing

The rubber pressing mold is the most basic and most common pieces of sole making equipment.
The rubber pressing mold is made of CNC cut or cast steel. Steel is used for rubber pressing molds due to is durability, low cost, and ability to retain heat.

The rubber pressing mold is essentially an over sized waffle mold. In fact, the first shoes made by Nike were called Waffle Racer due to their experiments making outsole parts with an actual waffle iron.

The function of the pressing mold is to form the raw rubber into shape while at the same time heating the rubber to vulcanize the compound. When you vulcanize rubber, you cook the rubber allowing it's internal molecules to cross link.

Raw rubber is soft and easy to tear, but once vulcanized it becomes the tough material we think of as rubber. The raw rubber is loaded into the mold in sheets. The heat of the steel and hydraulic pressure causes the rubber to flow into the mold's cavity displacing any air, then vulcanizes the rubber.

Once vulcanized, the rubber can not be remelted.

Simple Rubber Hot Pressing Mold

Cast steel hinged mold with Teflon coating.
Single color

Cost $800 to $1600 per size

This mold has a protective Teflon coating. You can see the conical alignment pins that will guide the mold halves into position as the press closes.

The deep groves around the perimeter of the mold cavity allow extra rubber and air to escape.

The molded parts look messy, but don't worry, the extra rubber is proof the entire cavity was filled and will be trimmed away.

The flat green
surface is the
middle plate

Middle Plate Rubber Pressing Mold

This rubber pressing mold operates on the same hot cycle as the single color mold, except there is a metal plate that divides the mold cavity into two compartments. This divider is called the middle plate.

These compartments can be the bottom tread and side wall. This is the most common type.

The classic Jordan 1 outsole is made this way. The dividing line is called the mold parting line or parting plane.

To produce the two-color sole, rubber is loaded into each compartment, with the middle plate installed to keep the two colors separate. The rubber is cured or vulcanized half way, the mold is then removed from the press. The operator then removes the middle plate and closes the mold allowing the rubber colors to fuse without running together.

These molds cost from $1200 to $1600 per size.

EVA forming equipment

EVA pressing equipment looks similar to rubber pressing but let's not confuse the two. The chemical processes are completely different.

We are going to describe the equipment needed to execute the EVA expansion and compression process. We will cover EVA injection tooling in Chapter 17.

It's important to note that Ethylene Vinyl Acetate (EVA) is a plastic material, NOT rubber. When we process EVA we are not vulcanizing the compound.

We must expand or "foam" the EVA, meaning we add a chemical to create air bubbles inside.

EVA foam expansion

EVA expansion molds are simple in design. Their function is to clear the EVA pre-form.

The EVA with foaming agent or blowing agent is loaded into the mold. These EVA pellets will expand and melt together.

The mold is bolted into a fast opening press. Once the EVA is heated, the press opens allowing the EVA to POP!

Just like popcorn! Tthe EVA is now filled with air bubbles. It is now foam.

Expansion mold
is 1/2 size

Fully expanded
EVA pre-forms

EVA pellets

Aluminum
EVA midsole mold

Simple EVA Hot Pressing Mold

Cast Aluminum Cast Steel with Teflon coating Heating and cooling cycles Cost $1200 to $1600 per size

The EVA mold is made of cast aluminum. The aluminum readily conducts heat. This is critical to the EVA pressing process.

EVA pressing is very different chemistry than that of rubber pressing. The EVA preform is fully cured material. The pressing operation heats and remelts the EVA material forcing it to conform to the mold cavity. The heat causes a slight expansion so the press is used to hold the mold closed.

After a heating cycle of several minutes, the EVA mold is shifted to a water cooled press bank to fix the final shape.

Once the EVA of removed from the cold mold, it is trimmed to remove any flashing, while the mold is reloaded and reheated.

The cycle time for heating and cooling can be 10 minutes per part. About twice the curing time of rubber.

Loading the pre-molded rubber parts

Combined Rubber and EVA Stock Fitting Molds

Cast Aluminum
Hot and cold cycles
Cost $1200 to $1600 per size

Assembling rubber and EVA parts together can be a time consuming and messy operation. This operation is called stock fitting. It's done on a short assembly line inside the molding factory or inside the shoe factory.

To save time, money, and to produce a better product, this stock fitting process can be dine "in-mold."

The rubber parts are molded with the addition of a layer of hot melt cement applied to the top surface. These rubber parts are then trimmed and inserted into the bottom of the EVA mold.

The EVA mold is designed with extra space to accommodate the rubber parts. The EVA blank is then inserted into the mold, making contact with the hot melt glue on the top surface of the rubber.

The EVA molding operation then proceeds as normal. Because the rubber parts are already vulcanized they are not effected by the heat required to melt the EVA. Vulcanized rubber can not be remelted.

Once the mold cools, the entire rubber and EVA part is removed, stock fitting complete.

Chapter 10

Cold Cast Resin Footwear Tooling

Cold cast or poured epoxy resin molds are often used to make Polyurethane foam midsole and outsole parts.

Resin molds are also used as low cost prototype tooling for new product development.

Liquid PU pouring

Cast resin molds for polyurethane
Cast resin molds are uniquely suitable for pouring polyurethane foam processing. The relatively low mold cost and high productivity of the typical polyurethane manufacturing line make this technology perfect for footbeds, midsoles, and other inexpensive parts.

Poured urethane does not require high clamping force, high heat, or high pressure, so cast resin molds are perfectly suitable for this process.

Liquid polyurethane has the ability to easily flow into complicated parts so it's easy to make high heel wedges with internal coring.

The low internal molding pressure and high flow of polyurethane makes this combination excellent for embedding airbags into the midsole.

On the next page you'll see an example of how the Nike Airmax midsole is made. The airbag is simply clipped into the mold and the PU is poured around it.

AIR MAX
AIR

The airbag is clipped into the mold

The cast resin mold is ideal for making high heel midsoles for wedge style shoes. You can see this design has significant coring inside. This coring saves weight and the low pressure of the molding operation allows us to insert reinforcement and anchor blocks.

In this case, a wood block is inserted into the mold and the PU flows around it securing it in place. The wood block becomes the nailing structure for the high heel footbed.

Liquid PU can flow to fill these complex shapes

This type of cast resin mold is ideal for footbed production. Low price and high production volumes are required for economical footbeds.

In this case, you can see the footbed top fabric is loaded into the mold and attached by pins. Once the mold is closed, the PU foams up and bonds to the fabric.

The parts are trimmed, and the footbed is complete.

This type of resin mold is also useful for prototyping midsole combinations. In this case, the outsole and midsole are combined into one piece. The inexpensive mold is created just for prototypes. For just a few hundred dollars, you can test the last fitting and upper design of a new shoe.

This type of mold can be made in just a few days. The prototype parts are inexpensive and flexible, and the PU is similar to EVA in its performance.

Once the design is confirmed the resin becomes obsolete, but the investment was quite low.

How are much resin molds?

The single color PU pouring mold is $400 to $500 per size.

The PU footbed mold costs $400 per mold, you may need 4.

Are these durable? Not so much.
Can you use the press EVA or rubber? No.

Can you use to make prototype test tools for development?
Yes you can.

Chapter 11

Temporary Prototype Tooling

The metal casting and cutting operations we have seen thus far are expensive and time consuming.

How else can we make a working prototype?

Silicone rubber, cold cast resin, and 3D printing are all viable options.

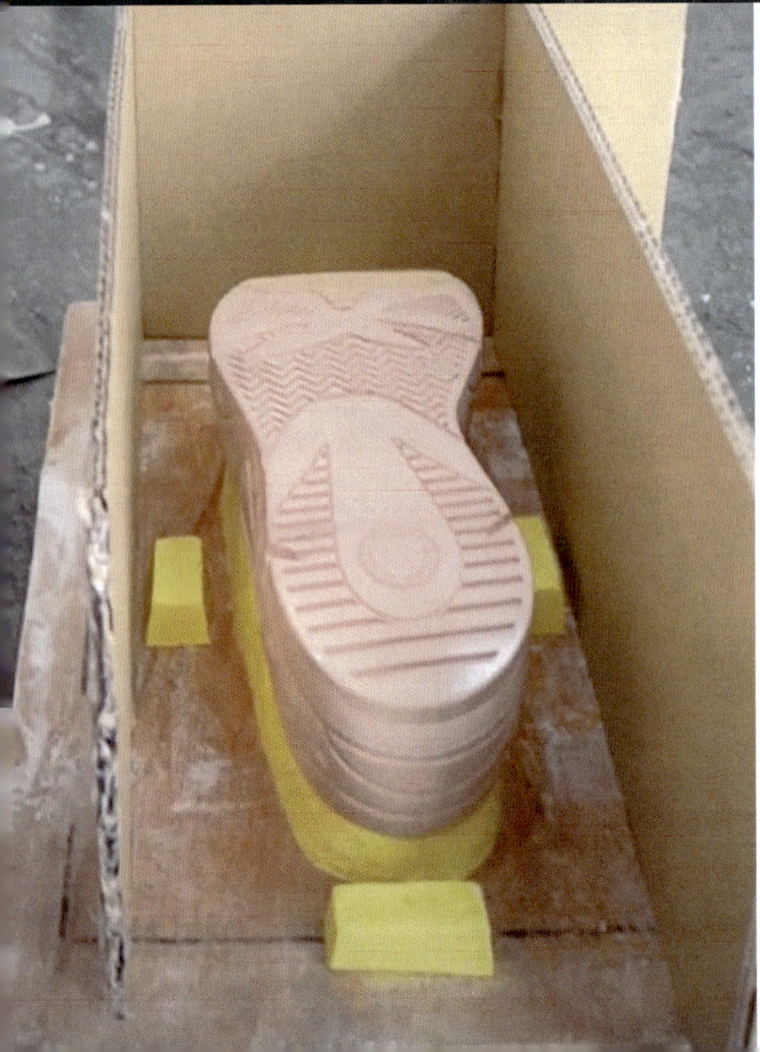

Making a prototype tool

The process for making a silicone mold is pretty straightforward. The wood cut used to confirm the design is covered with mold release and packed with crafting clay.

The crafting clay fills up the entire top net of the outsole unit. The crafting clay is then secured to the bottom of a temporary mold box.

The silicone rubber compound is simply poured in on top of the wood cut and clay.

Once the silicone is cured, the clay is removed exposing the top net of the outsole unit. Once the clay is removed, the silicone mold and top net are treated with mold release.

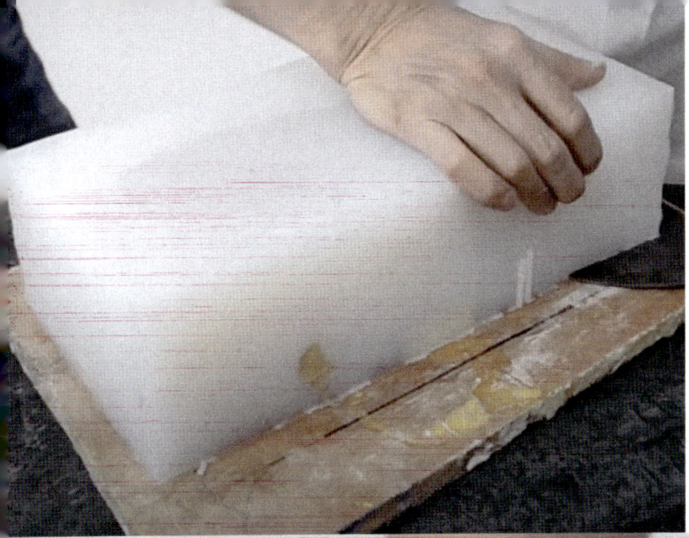

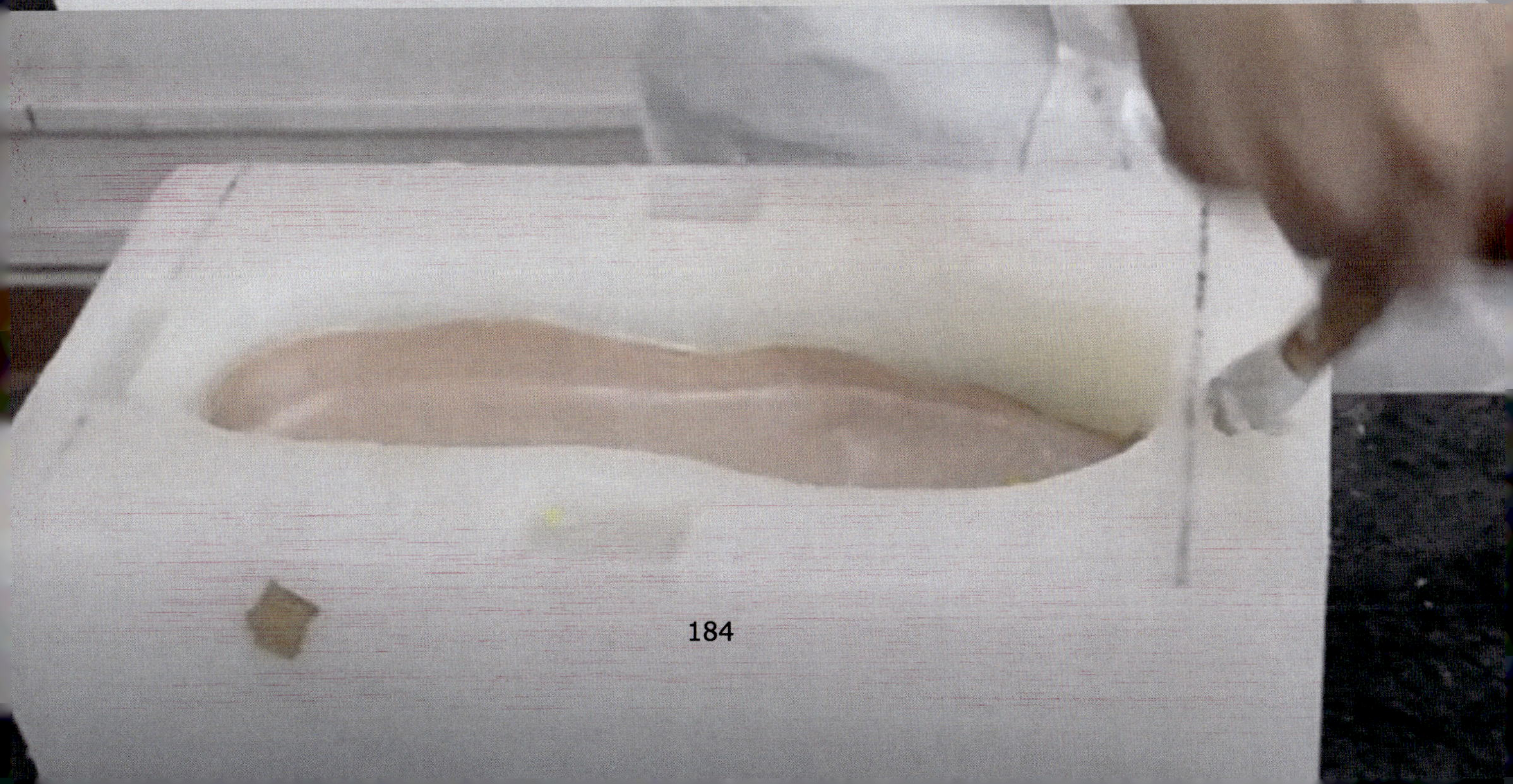

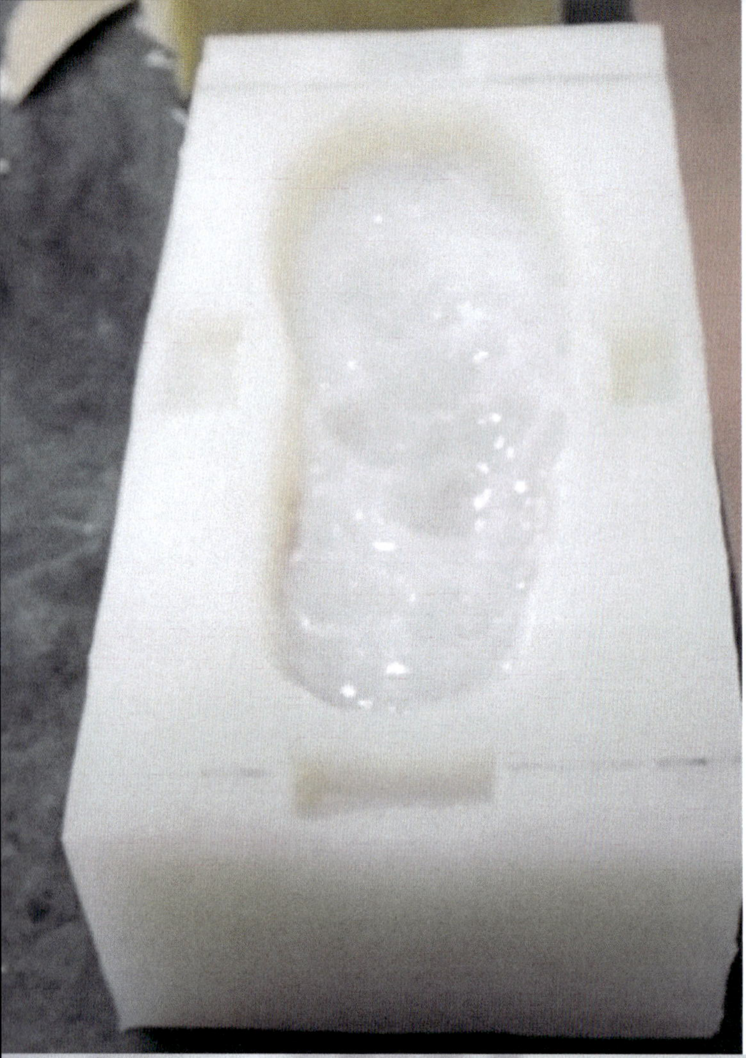

Compound is then poured into the top cavity.

Once the silicone is cured, the top plug can be removed, exposing the wood cut. Because the entire mold is flexible, the silicone can be peeled back from the wood cut, exposing the complete cavity.

This type of mold can be made for a few hundred dollars and can be done almost overnight.

The silicone mold is useful for 10 to 15 impressions.

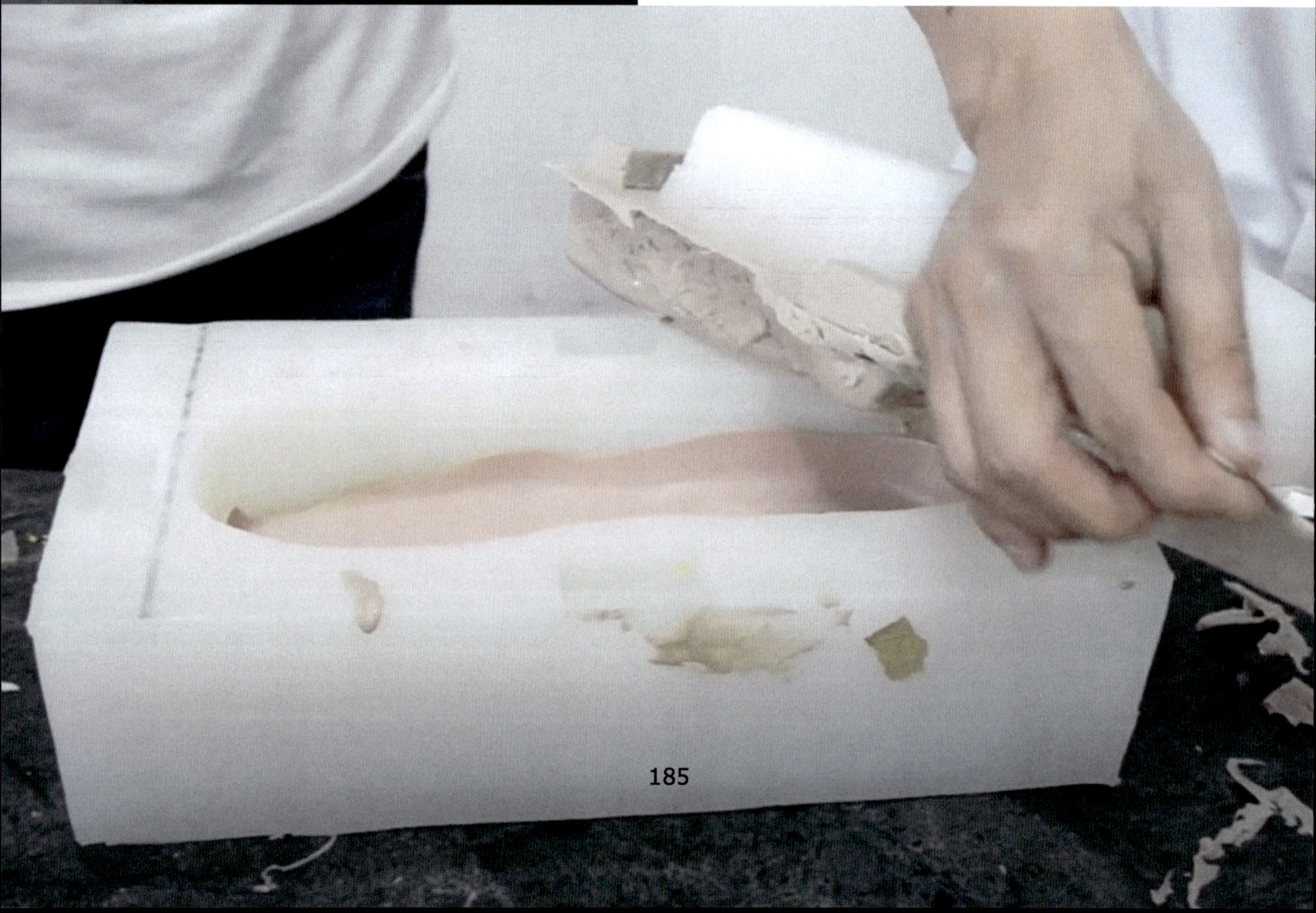

31
31

-B

As we discussed in the previous chapter, cast resin can also be used for this type of prototype mold.

When making a prototype part, the wood cut is created with the rubber and midsole parts together so they can move as one piece.

This type of resin prototype mold costs a little bit more money, takes a little bit longer to make, but it is useful for many hundreds of parts.

These urethane parts can be painted to look like the real components. Because these parts are made of urethane foam, which is a typical midsole material, test shoes made with this prototype technique are suitable for wear testing, marketing, videos, and photos shoots.

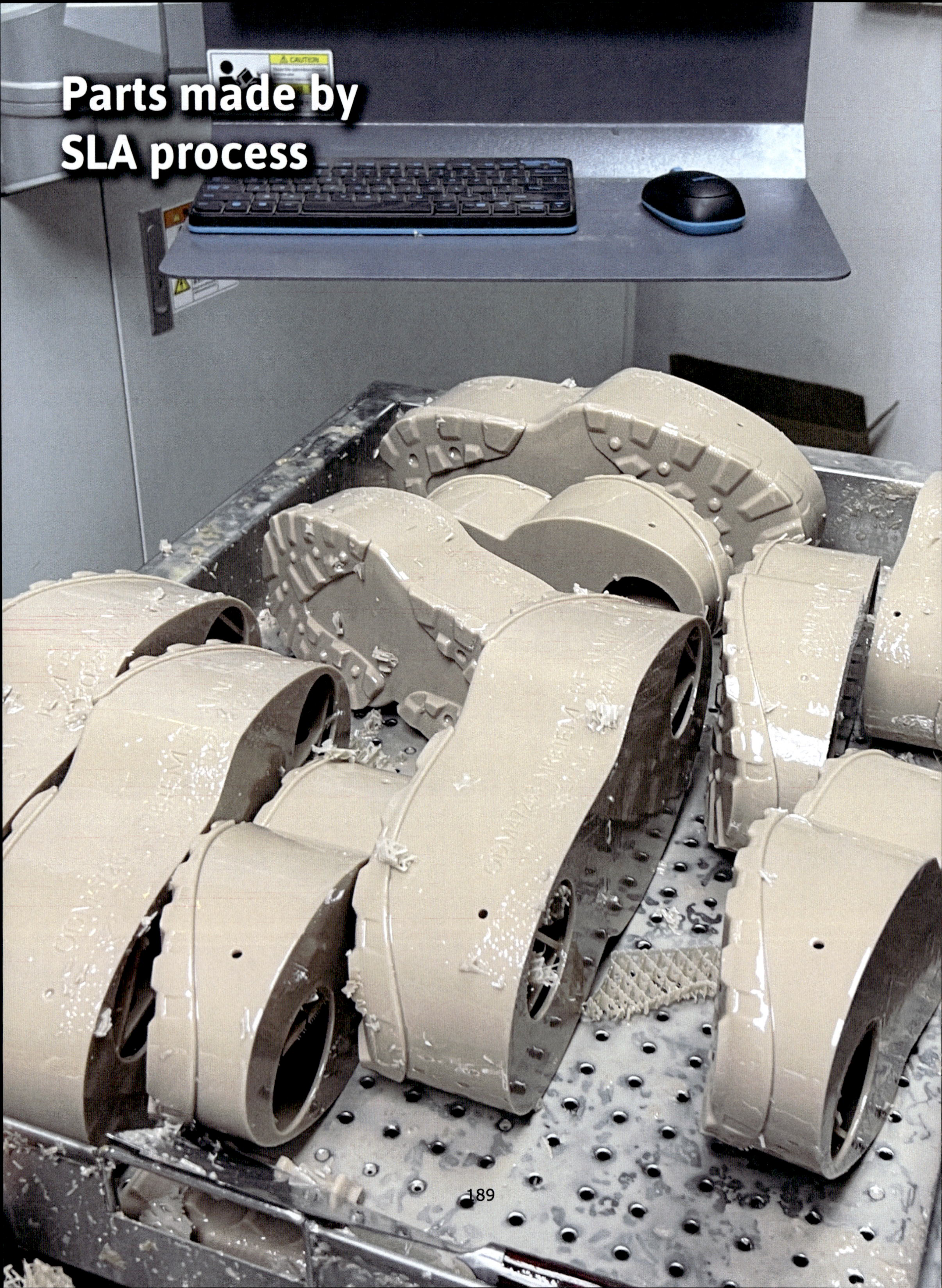

Parts made by SLA process

3D Printing Prototypes

The same 3-D printing machines we use to create wood cuts can also be used to create cosmetic models for product photography. These 3D molds cannot be worn, but they will be much more precise than the cast silicone or cast resin molds.

3D printed plastic molds

Yes, the 3-D printing process can be used to make temporary plastic molds. This technology is relatively new and rare.

Chapter 12

3D Printed Outsole Tooling

A new technology with new features.

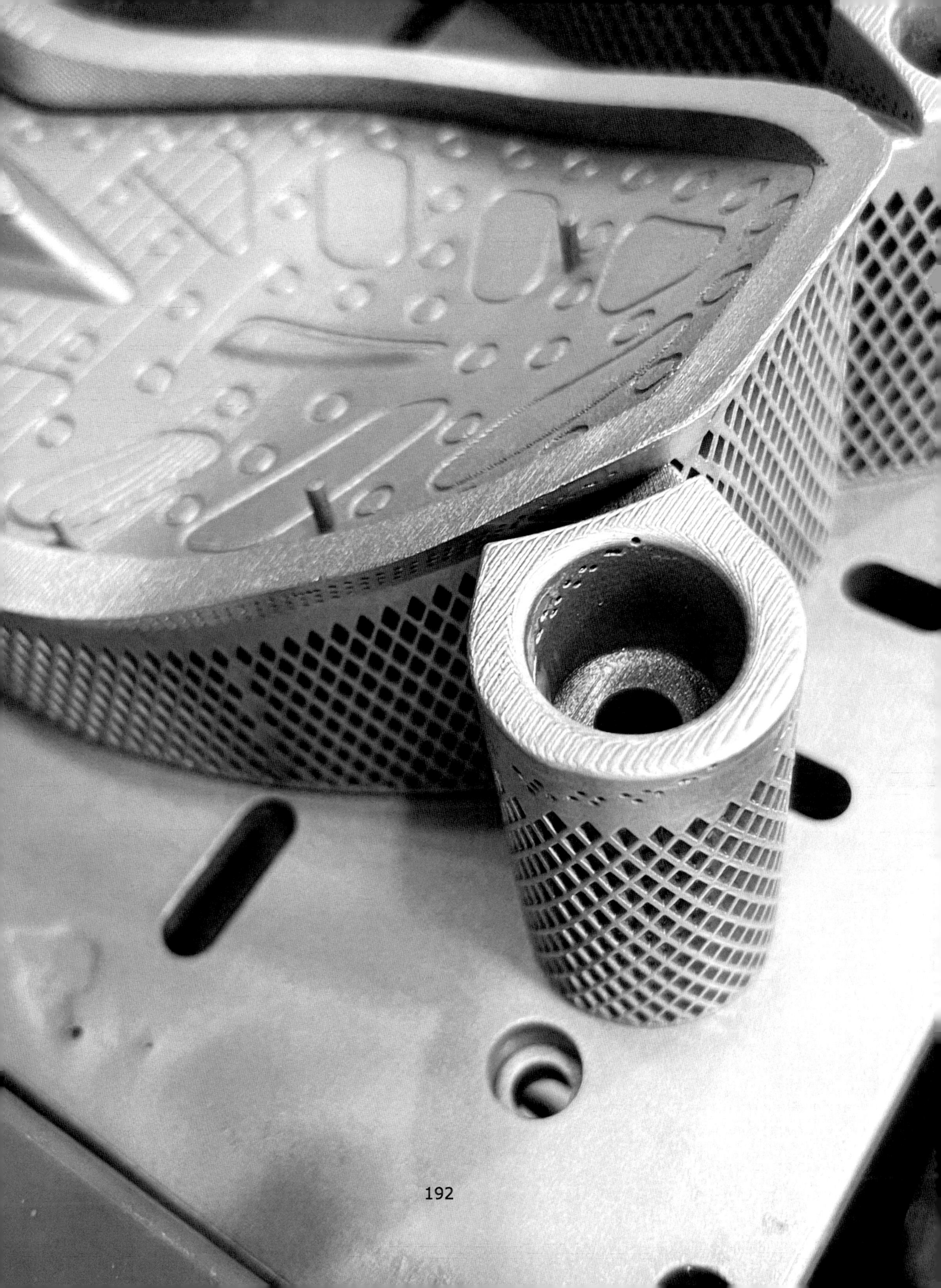

3D Printing Metal Molds for Shoe Sole Production

The footwear industry is rapidly adopting 3D printing technology to enhance manufacturing efficiency and design flexibility.

One of the most significant advancements in this space is the use of 3D printing to produce metal molds for shoe sole production. Traditional mold-making methods, such as CNC machining, are time-consuming and expensive, but metal 3D printing offers a faster and more cost-effective alternative.

Using technologies like Direct Metal Laser Sintering (DMLS) or Selective Laser Melting (SLM), manufacturers can create high-precision metal molds with intricate tread patterns and customized designs. These advanced techniques allow for the production of complex geometries that would be difficult or impossible to achieve with conventional machining. Additionally, 3D printing reduces material waste and speeds up the prototyping process, enabling designers to quickly test and refine their sole patterns before mass production.

This type of 3D printed mold is used quite commonly for E-TPU midsole production. The 3-D printing process allows very unique air venting through the sidewall of the mold. This allows the boost type to be made without the signature boost popcorn texture.

The entire sidewall of the mold is covered with micro vents.

These 3D printed molds allow you to make geometry that's impossible for a cast or CNC cut mold. For example, rubber lugs with opposing draft angles are impossible, except with this new mold making process.

This 3-D printing process allows for texture designs that are deeper and detailed to a higher level not possible with traditional mold texture techniques.

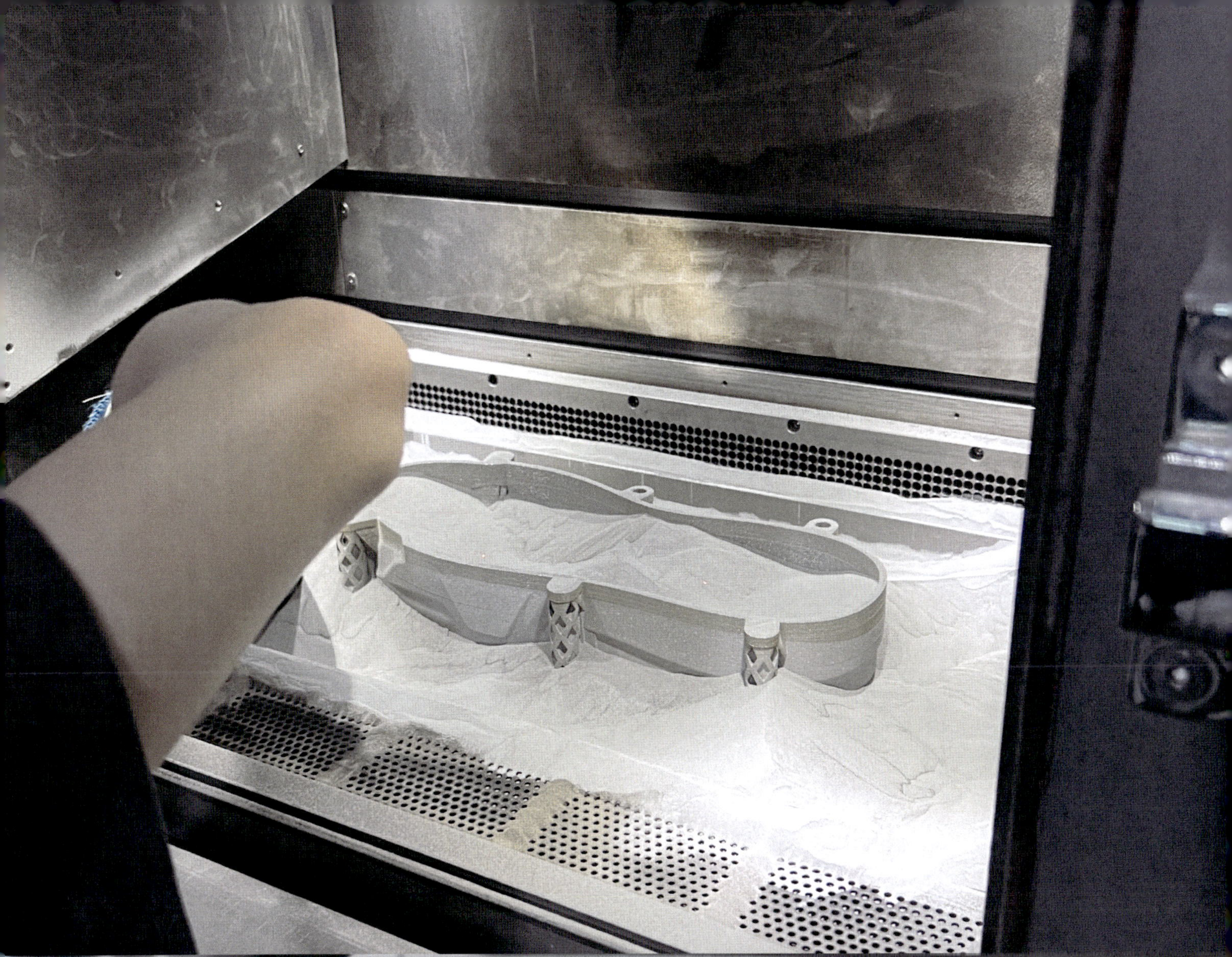

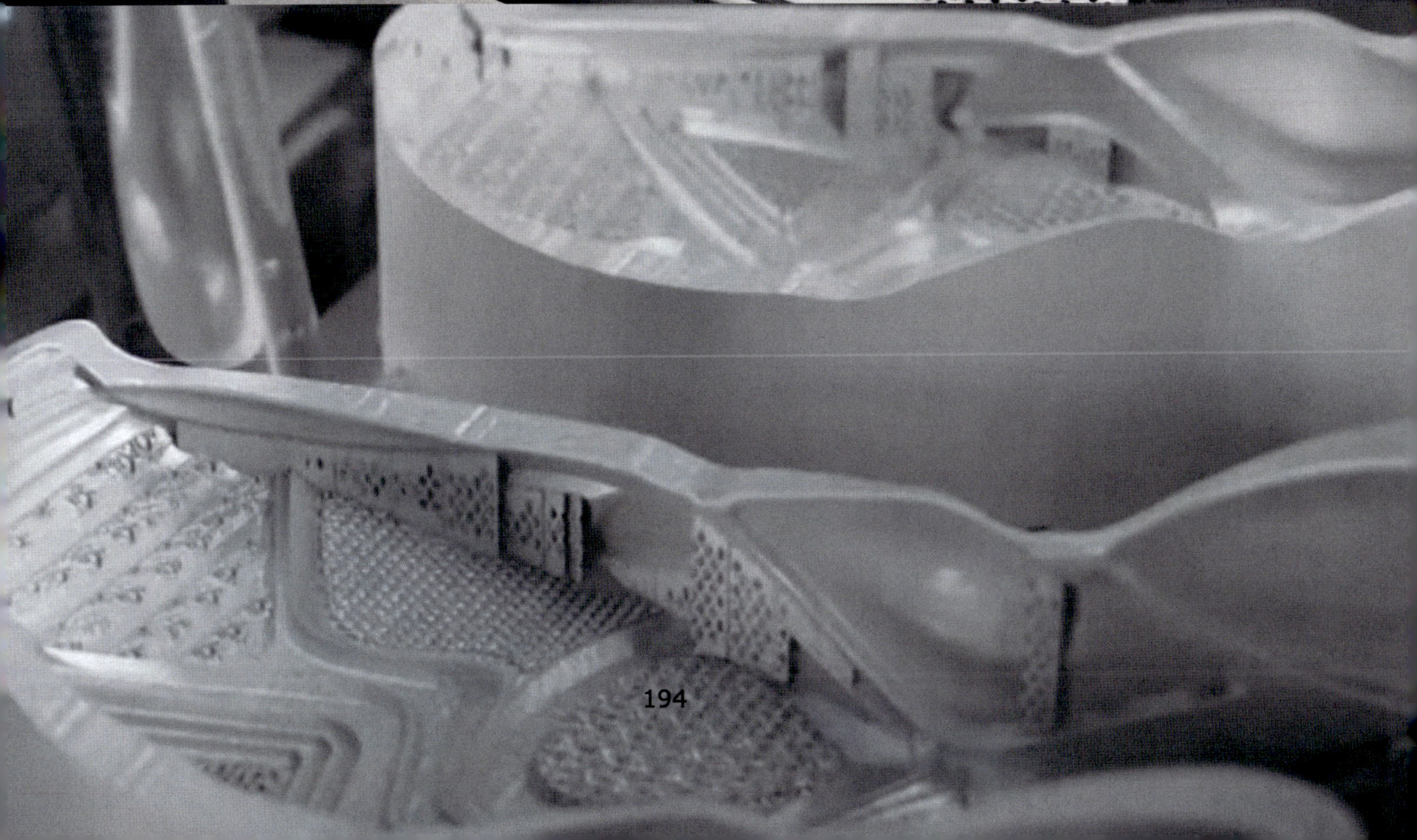

Metal
3D
Printing

How much are 3D printed molds?

These are rare. Not may mold shops have this equipment. Expect to pay double the cost of a cast or CNC cut mold.

These can be $2500 to $3000 per size.

RISCALDAMENTO
FORME
SUOLA

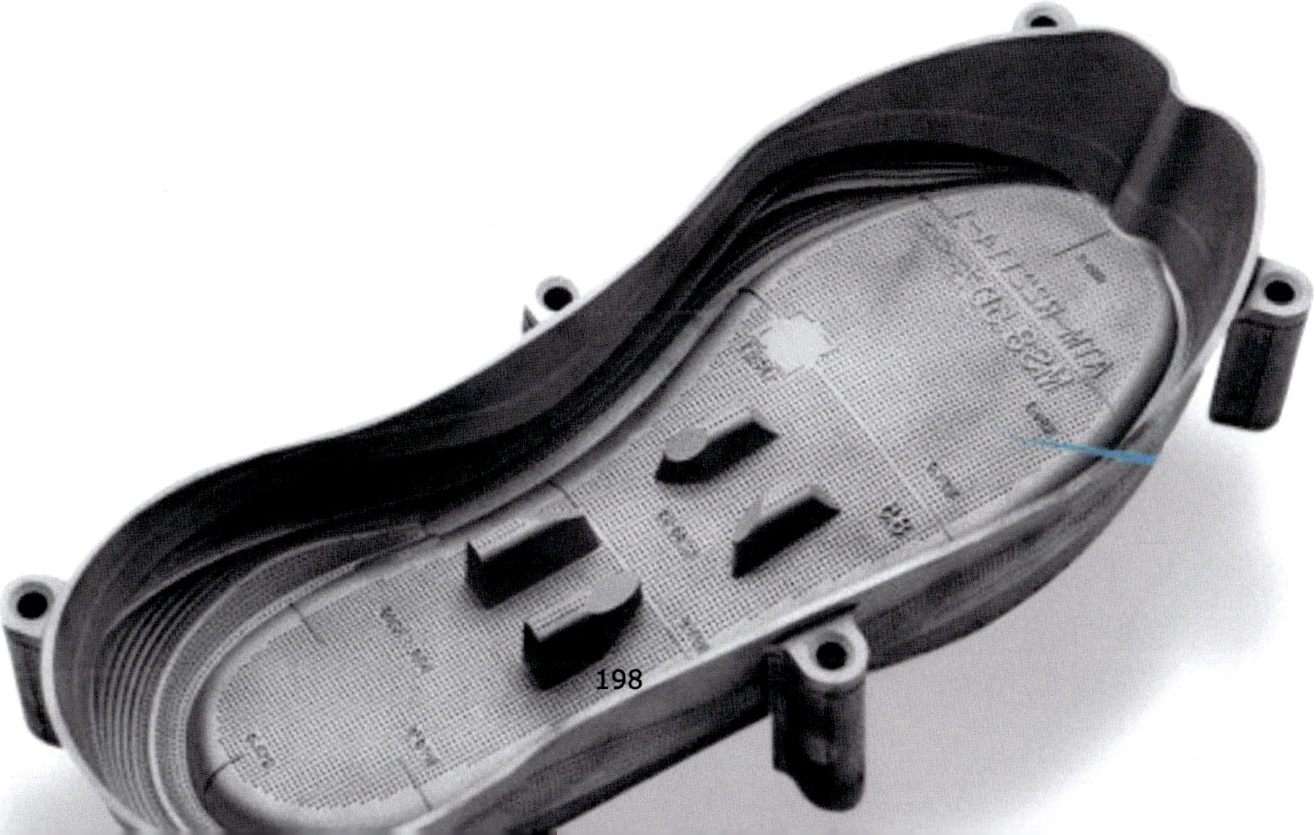

Chapter 13

Plastic Injection Molds For Footwear

Plastic injection molding is what makes modern sport shoes really modern.

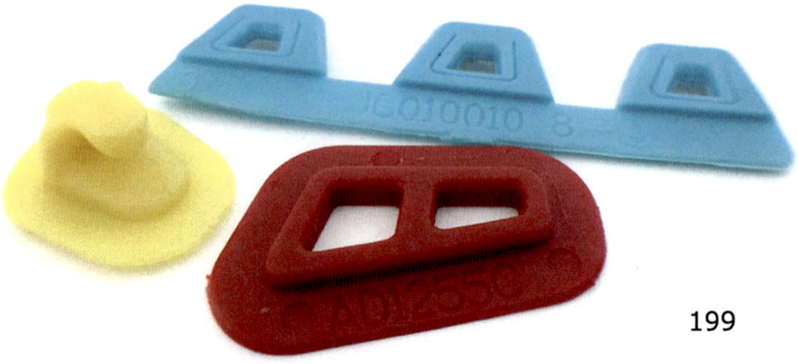

Injection Molding in Footwear Manufacturing

Injection molding is a highly specialized manufacturing process with extensive applications across various industries.

While one could dedicate an entire academic career to studying this technology, this overview focuses on its relevance in footwear production.

Overview of Injection Molding

Injection molding is a process in which heated plastic is injected into a mold under high pressure, forming a specific shape upon cooling and solidification. This technique is widely used in the production of footwear components offering precision, durability, and cost-effectiveness.

Injection Molding in Footwear

Footwear manufacturers utilize injection molding for various components including outsoles, midsoles, cleats, and structural reinforcements. Some notable examples include:

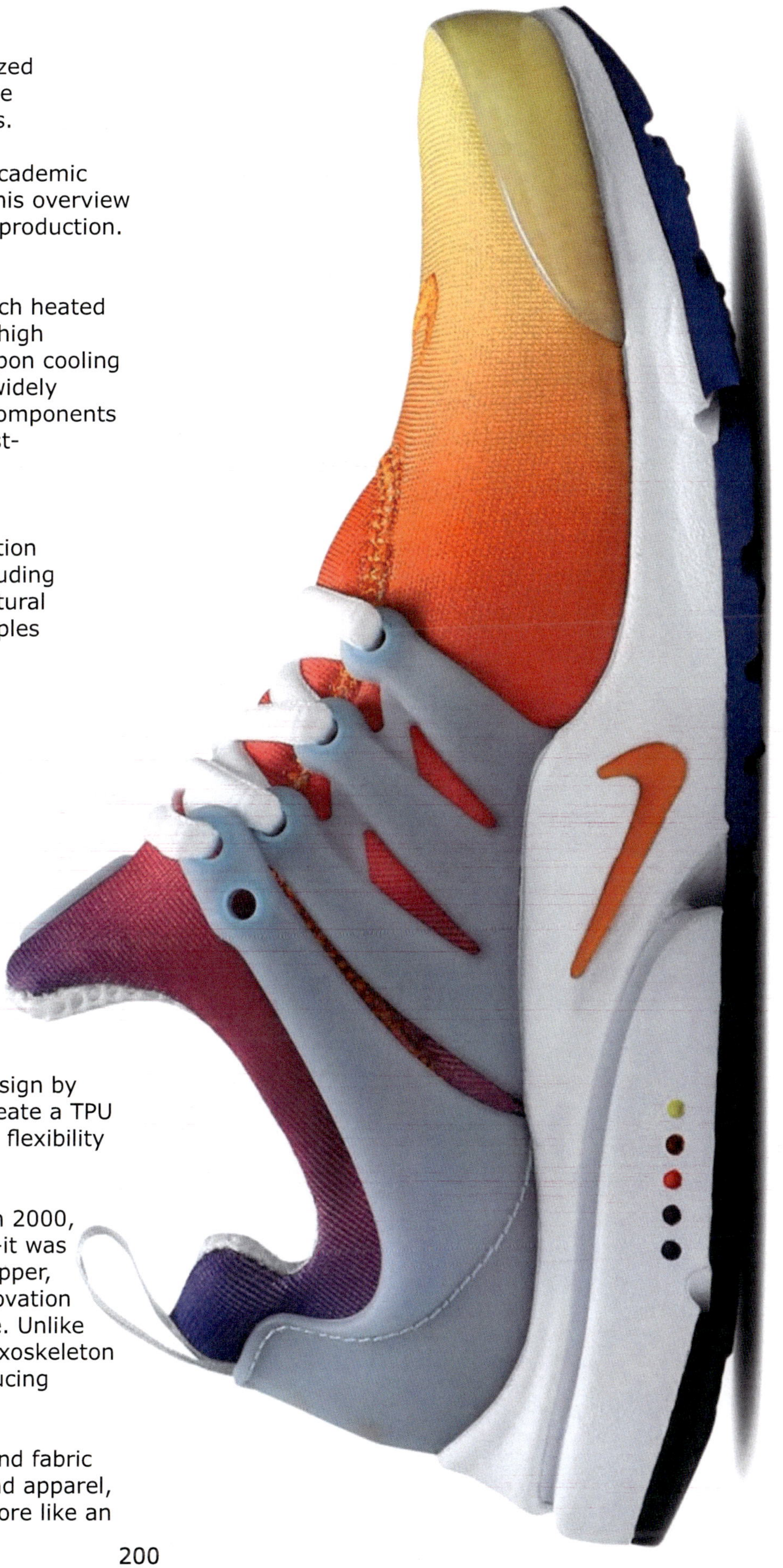

Nike Presto

This shoe revolutionized footwear design by incorporating injection molding to create a TPU mold reinforced with nylon, ensuring flexibility and durability.

When Nike launched the Air Presto in 2000, it wasn't just another running shoe—it was a revolution. Its stretchy, sock-like upper, delivered a snug fit, but the real innovation is in its injection-molded plastic cage. Unlike traditional eyelets, this translucent exoskeleton provided adaptive support while reducing weight.

The seamless integration of plastic and fabric blurred the line between footwear and apparel, making it feel less like a shoe and more like an extension of the body.

A futuristic design ahead of its time.
The Presto was not the first....

The Puma King 1966
In 1966 Puma brought injection technology to the soccer shoe. Adidas soon followed. From that moment on, plastic became the material of choice for soccer cleats.

Modern soccer cleats often feature multiple injection-molded compounds, enhancing performance and aesthetic appeal.

Performance Footwear
Injection molding is also integral to products like baseball cleats, cycling shoes, ice skates, snowboarding boots, and wakeboard bindings, all of witch require durable, precision-engineered components.

Design Flexibility
Injection of plastic is an incredible material for footwear projects. It can be rubber soft or it can be rock hard. It can be water clear. It can be shiny or matt finish. Plastic parts can be multi density and multicolor.

We can add carbon fiber or glass fill to make plastic lighter and stronger. There are almost no limits to the colors. Plastic injection parts will hold their detail. We can put fine, molded textures, and shiny finishes all on one part.

Plastic parts can also be painted and clear parts can be back printed.

Injection molds are also highly productive.
The cycle time for plastic parts is quite short, less than a minute in most cases. Part prices, depending on the material specification, can be quite reasonable.

The only downside is the expense of the molds. You will see that plastic molds can cost from a few thousand dollars to many tens of thousands of dollars.

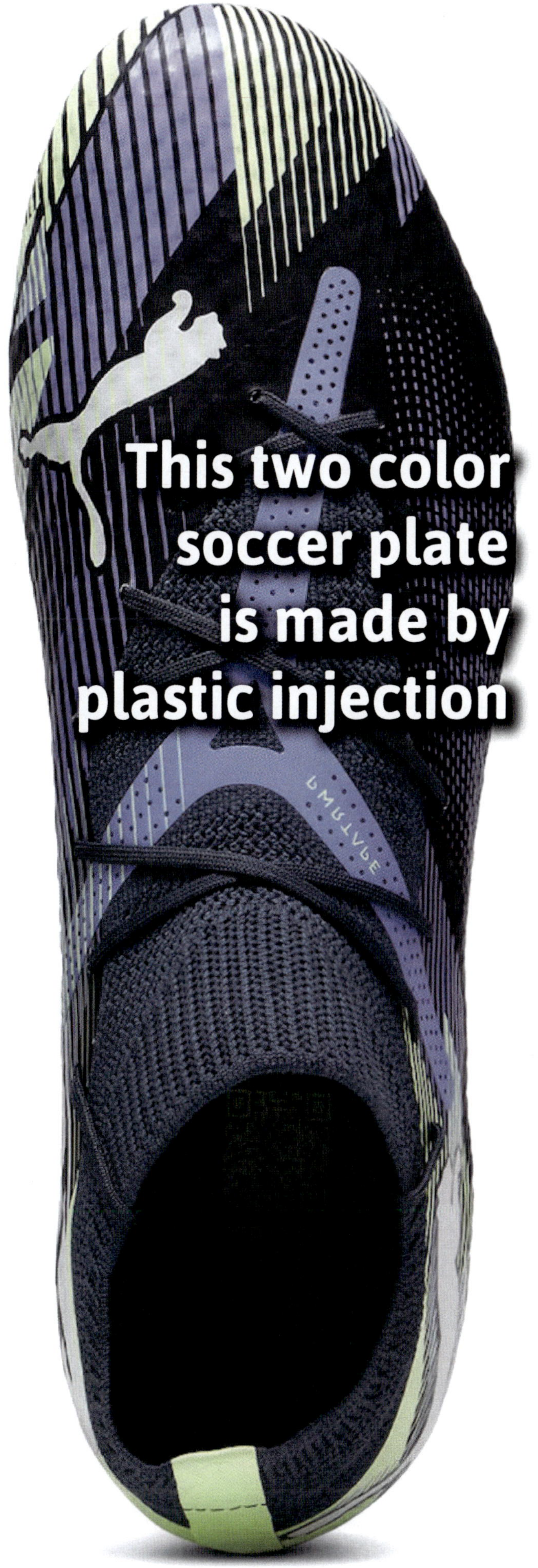
This two color
soccer plate
is made by
plastic injection

A typical injection shop making footwear parts

The Injection Molding Process

The process involves several key steps:

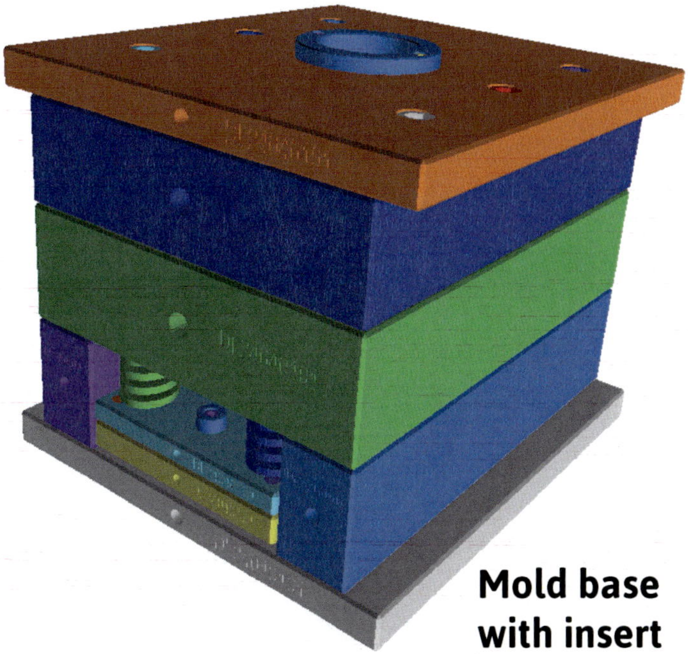

Mold base with insert

Material Preparation:
Plastic pellets are heated until they reach a molten state.

Clamping Phase:
Powerful hydraulic rams close the mold and hold it shut. It requires many tons of force to hold the mold closed while the plastic is injected under high pressure.

Injection Phase:
The molten plastic is forced into the mold cavity under high pressure using a screw-drive mechanism. The injection molding technicians carefully adjust the machine with the right amount of force, the right amount of plastic, the right temperature, and the right speed, to allow the plastic to flow into the mold. As the plastic flows in, the air must flow out. Injection molding parameters are very complicated. A new mold may take several days to get all the parameters adjusted correctly.

Cooling and Solidification:
The injected material cools and hardens to take the desired shape. The technicians job is not over once the plastic is inside the mold. The machine will require more adjustments for how long to hold the pressure and at what temperature. Injection molds will have cooling lines with water to help speed up the cycle time. Only after the plastic has set can the mold be opened.

Ejection:
The final part is removed from the mold, often requiring additional trimming or finishing. Once the part is ready, the mold will open and injector pans will push the plastic out of the mold cavity. Imagine how the plastic fits the cavity perfectly so it takes mechanical aids to kick the plastic out of the tool. Injected parts may have pin marks but a properly designed tool will hide those. Softer components may have to be peeled out of the mold.

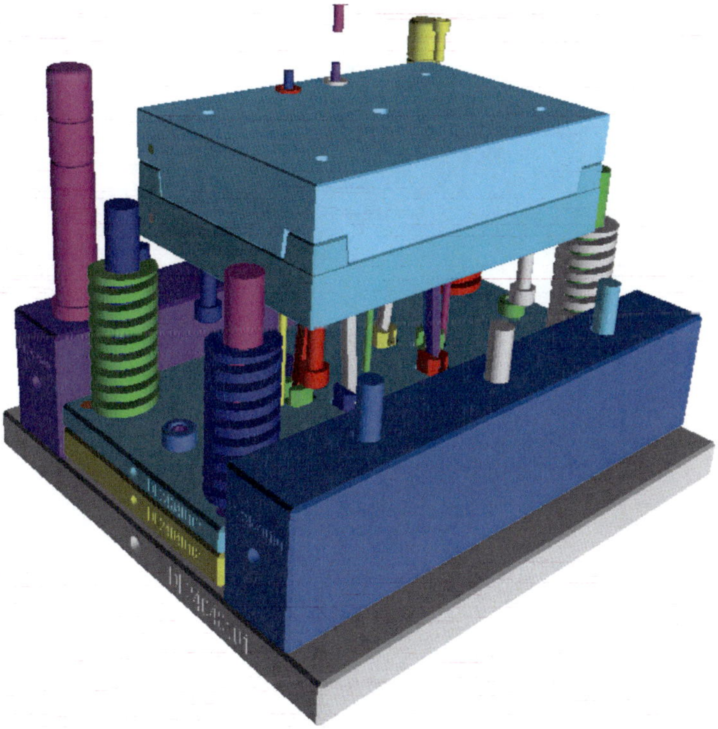

Injection molding requires substantial force to ensure the plastic properly fills the mold cavity. This force is measured in tons with machines ranging from 110 tons to over 1,000 tons depending on the complexity of the mold.

The size of the machine is measured in tons of clamping force.

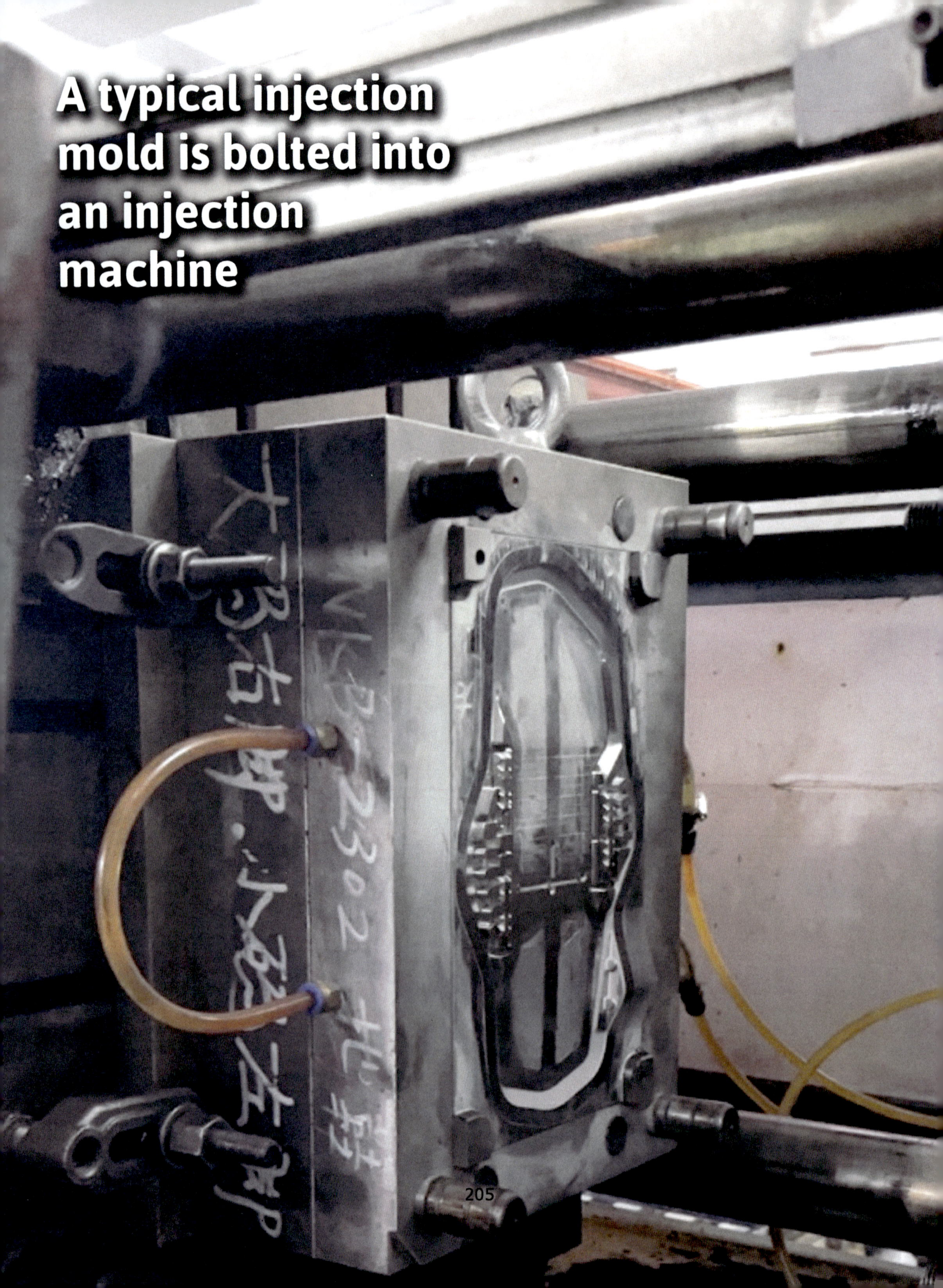

A typical injection mold is bolted into an injection machine

Machine Size
The clamping force is crucial because it keeps the mold tightly closed during injection to prevent flash (excess plastic escaping) and ensure proper part formation. The required force depends on the part size, material type, and cavity pressure inside the mold.

Small machines: 30–100 tons (suitable for small plastic components like medical devices, electronics, or shoe parts.)

For something like a shoe sole or a cleat bottom, a machine in the 50–150 ton range might be used, depending on material flow and part complexity.

A plastic curb size garbage bin requires a 3000 ton machine!

Types of Injection Molding in Footwear

Single-Color and Multi-Color Molding:
Some products, like soccer cleats, require multiple colors or materials, achieved by either sequential injections or over-molding techniques.

Rigid and Flexible Components: Injection molding can produce rigid parts (e.g., ski boot shells) or flexible elements (e.g., sneaker midsoles and outsoles).

Direct Injection Molding:
This technique fuses the outsole directly onto the upper, commonly seen in PVC rain boots and work shoes.

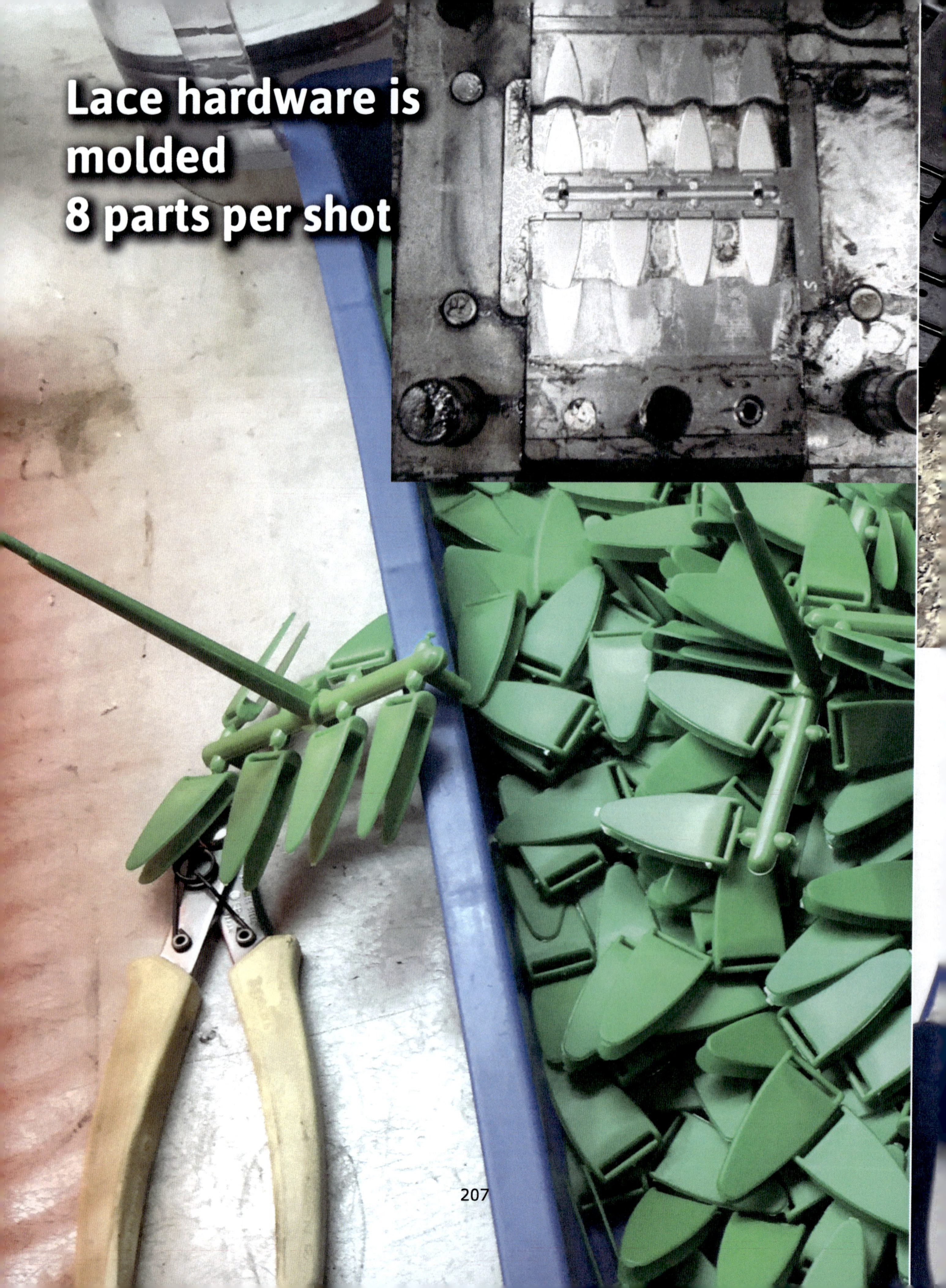
Lace hardware is molded
8 parts per shot

Glass filled
nylon shank

High-heels require strong ABS plastic cores

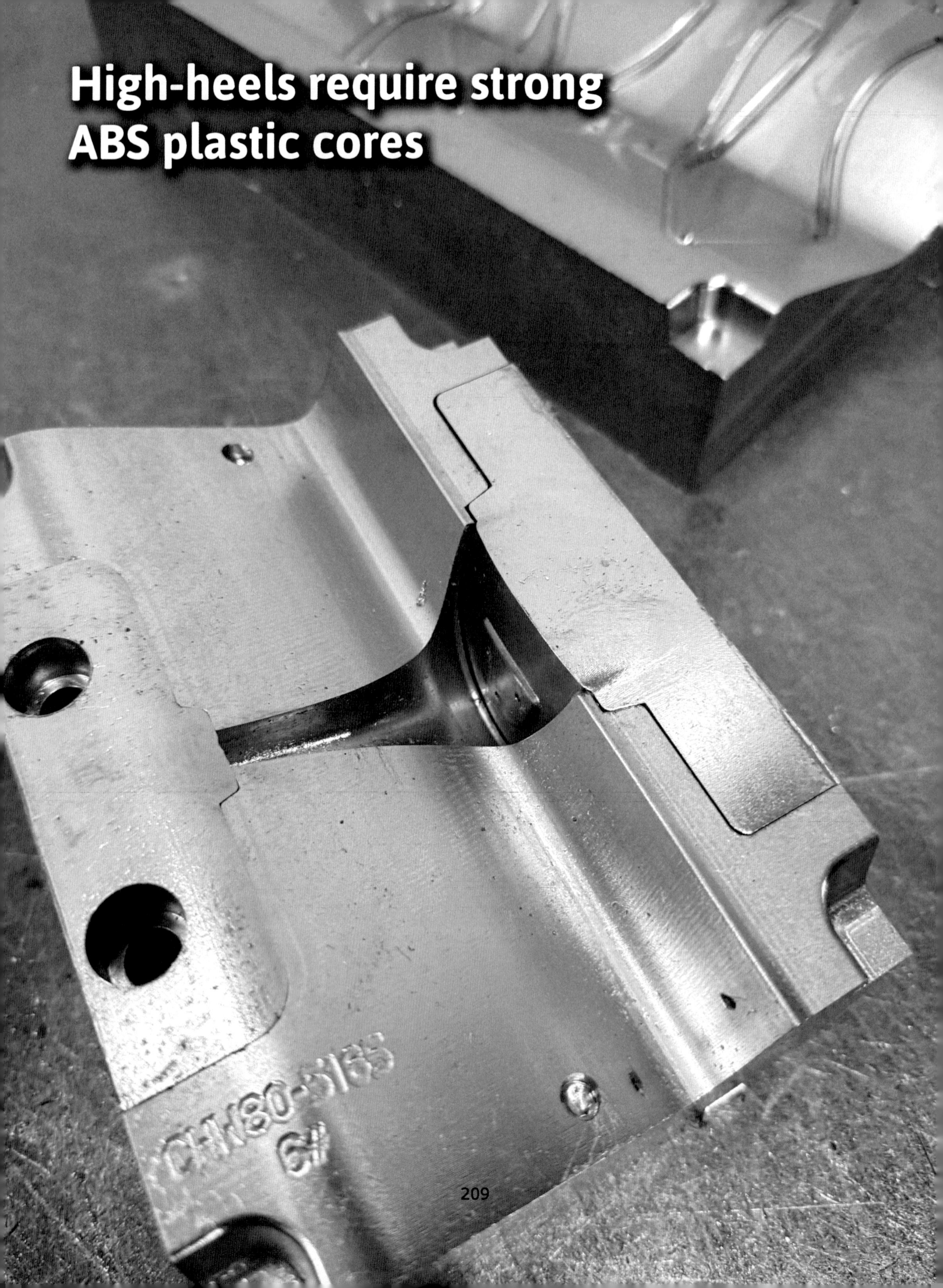

Flexible TPU upper molds with over molded nylon lace inserts

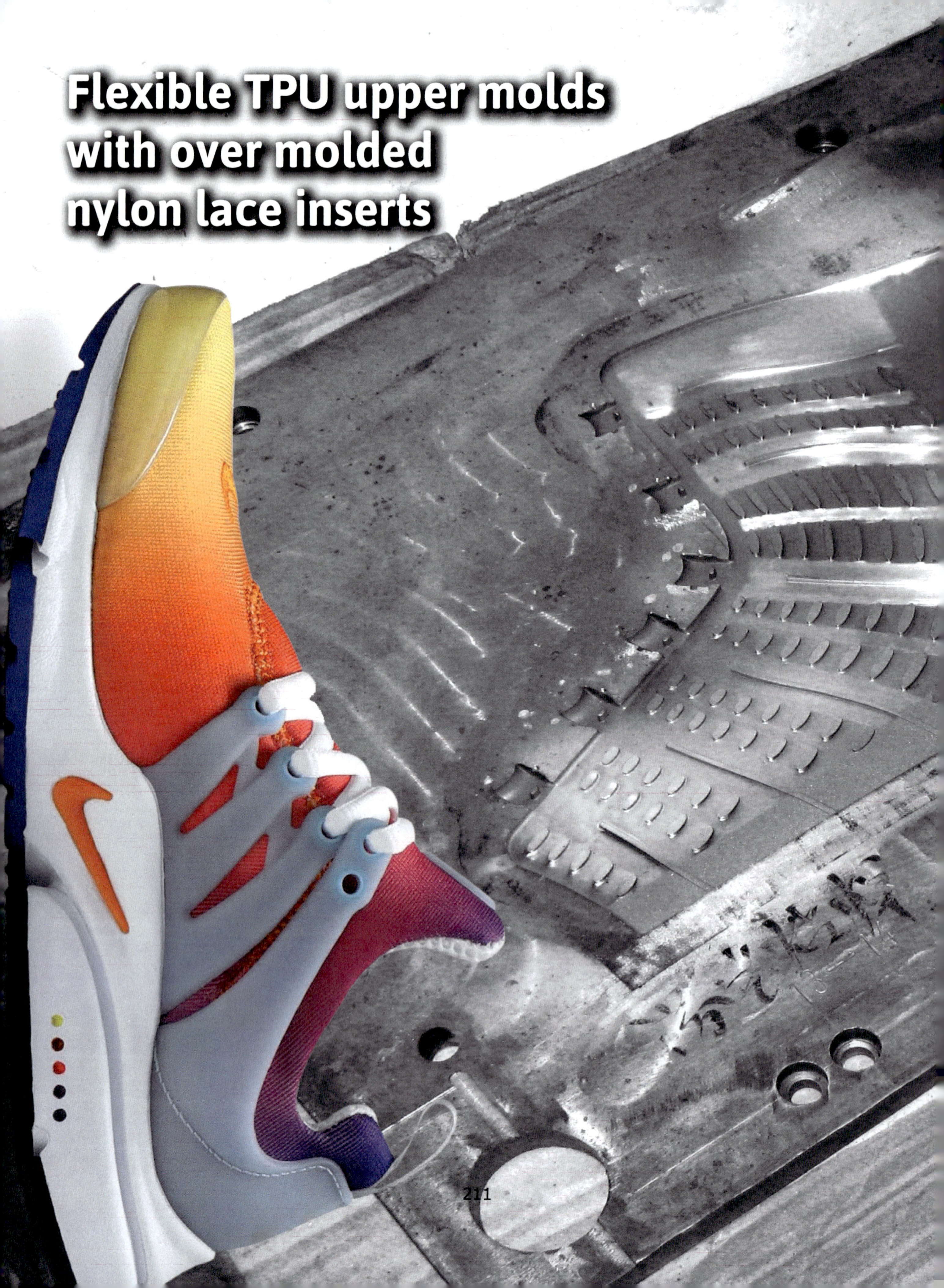

CONTROL

Over molding or insert molding

In these examples fabric is placed directly into the injection mold. When the mold closes, the plastic is injected fusing itself directly to the face of the fabric.

This is a very challenging mold as you can see the plastic must flow around all those little islands to connect together to make the squares and hexes.

So this looks like a simple mold, but requires high pressure so the plastic can fill all those tiny spaces.

PVC rain boots made by overmolding a polyester sock

Larger parts with more complicated designs will require molds that have inserts or open with angled slides.

In this example mold, The mold components slide away on angled shafts releasing the complicated shape from the mold. This type of mold can be very expensive. Depending on the size 5 to $10,000 per mold, per left and right Per size.

You can imagine how quickly injection mold costs add up. I'm like this example may be good for half 1 million shots or more.

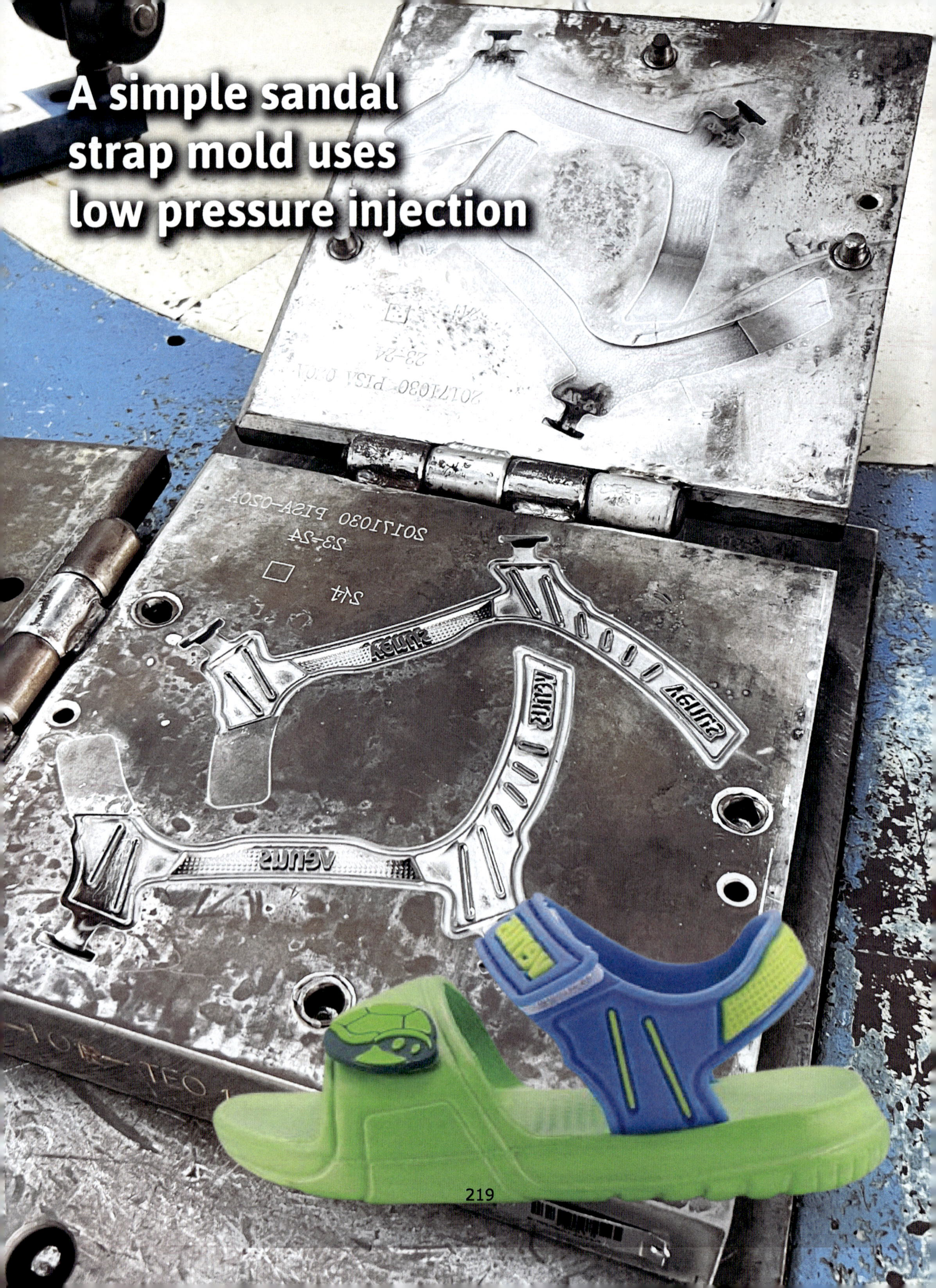
A simple sandal strap mold uses low pressure injection

A ski boot
requires an expensive
high pressure
mold
NORDICA
$45,000 per mold

The most common plastics used for injection

Thermoplastic Polyurethane (TPU)
Used for: Outsoles, midsoles, cleats, support cages, and overlays.
Why?: Durable, flexible, and abrasion-resistant, making it perfect for high-performance sports and outdoor footwear.

Nylon (PA)
Used for: Structural reinforcements, upper components, and plates (e.g., cleat plates).
Why?: Strong, flexible, and lightweight, providing excellent support in performance footwear.

Polyethylene (PE)
Used for: Lightweight outsoles, waterproof layers, and reinforcement elements.
Why?: Low-cost, lightweight, and moisture-resistant, commonly found in disposable and casual footwear.

Polyvinyl Chloride (PVC)
Used for: Cheap outsoles, rain boots, and casual shoes.
Why?: Inexpensive, flexible, and waterproof, though less eco-friendly than some alternatives.

Thermoplastic Rubber (TPR)
Used for: Outsoles, grip-enhancing components, and flexible sections.
Why?: Combines the flexibility of rubber with the ease of injection molding, making it ideal for slip-resistant and comfortable footwear.

Acrylonitrile Butadiene Styrene (ABS)
Used for: Hard protective components, heel counters, midsoles in specialized footwear.
Why?: Strong, impact-resistant, and rigid, often used for added support or protective elements in work boots and sports shoes.

Basic Molds: Small injection molds start around $2,500.

Advanced Molds: Large or complex molds, such as those used for ski boots, can cost upwards of $45,000 per size.

Advantages of Injection Molding
Precision: Ensures consistent and accurate component production.

Durability:
High-quality materials and strong molds result in long-lasting parts.

Efficiency: The process allows for high-volume production with minimal waste.

Versatility: Supports a variety of materials, from rigid plastics to soft elastomer.

The final word on plastic molds
Injection molding is an essential process in modern footwear manufacturing, offering both design flexibility and production efficiency.

It enables the creation of complex, high-performance footwear components while maintaining cost-effectiveness. By understanding the various injection molding techniques and their applications, manufacturers can optimize product design and production efficiency, ultimately delivering superior footwear to consumers.

The injection molding process can be used to make many different shoe parts.

Many different materials can be injection molded.

We can make soft, gel-like parts, tough, flexible eyestays, and stiff shanks all by injection molding. Regardless of the material, the mold making process is about the same.

Chapter 14

PVC & TPR Outsole Injection Molds

Easy to mold, relatively inexpensive, great for cheap shoes and kids shoes.

But, a little bit heavy for adult sized footwear.

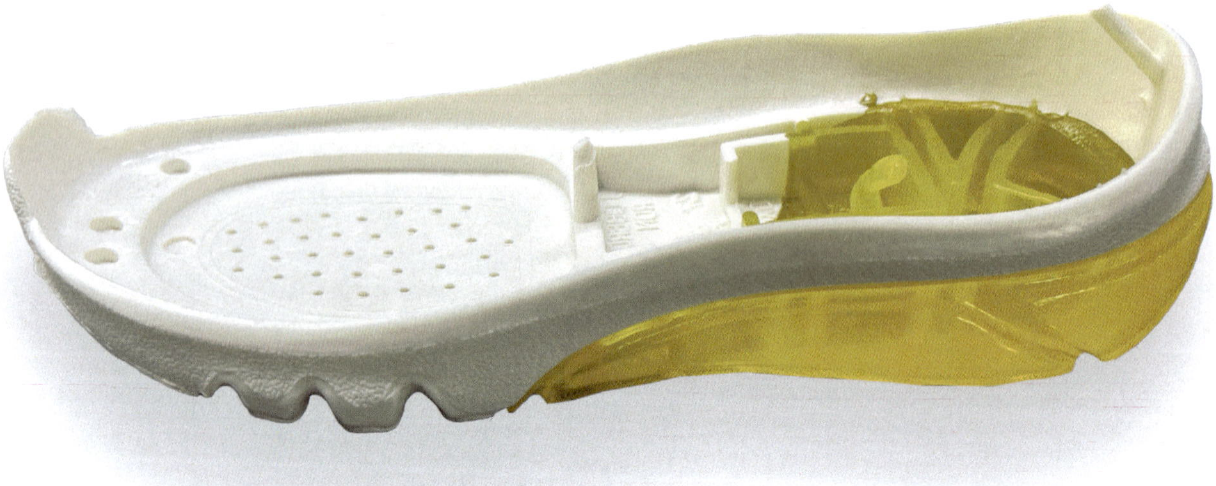

Two color
rotating mold

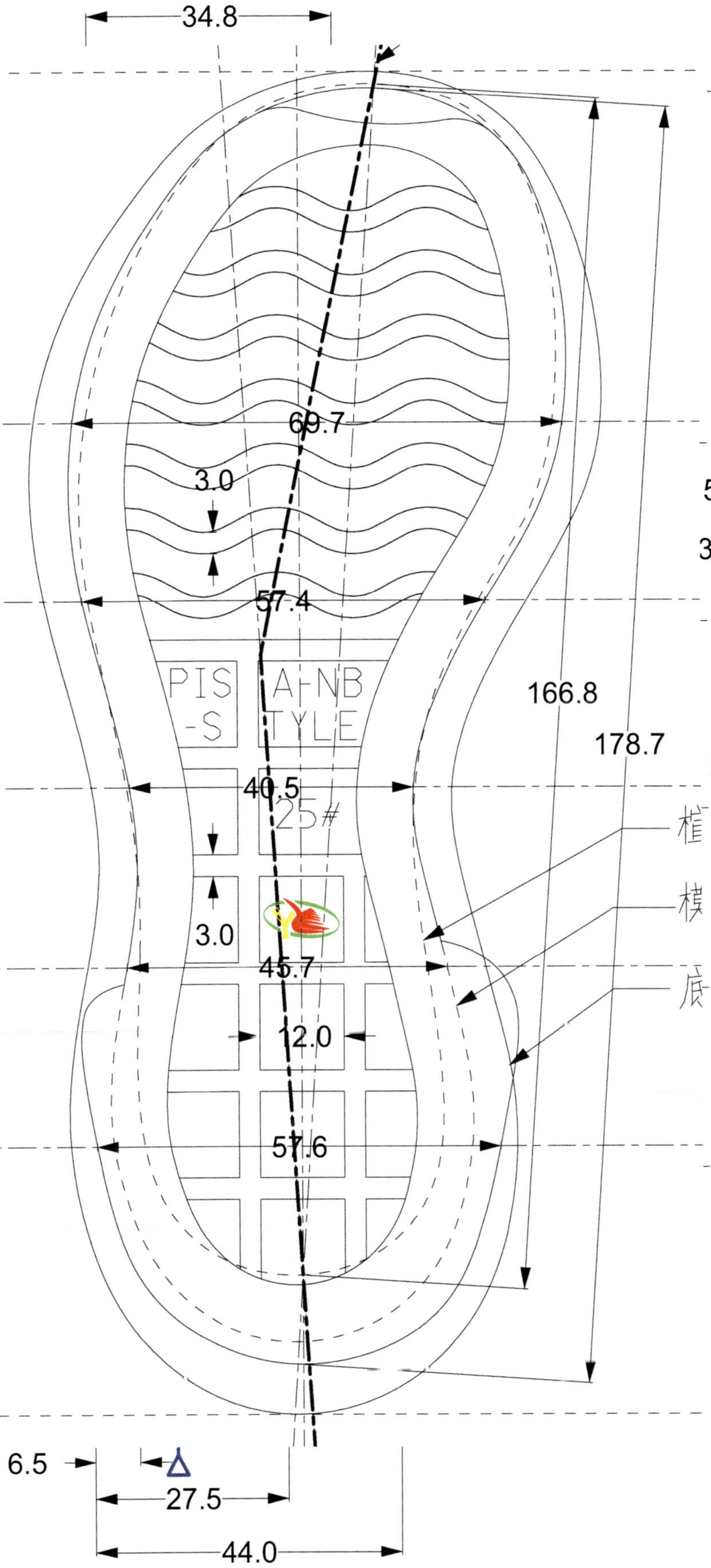

LATERAL VIEW

外侧图

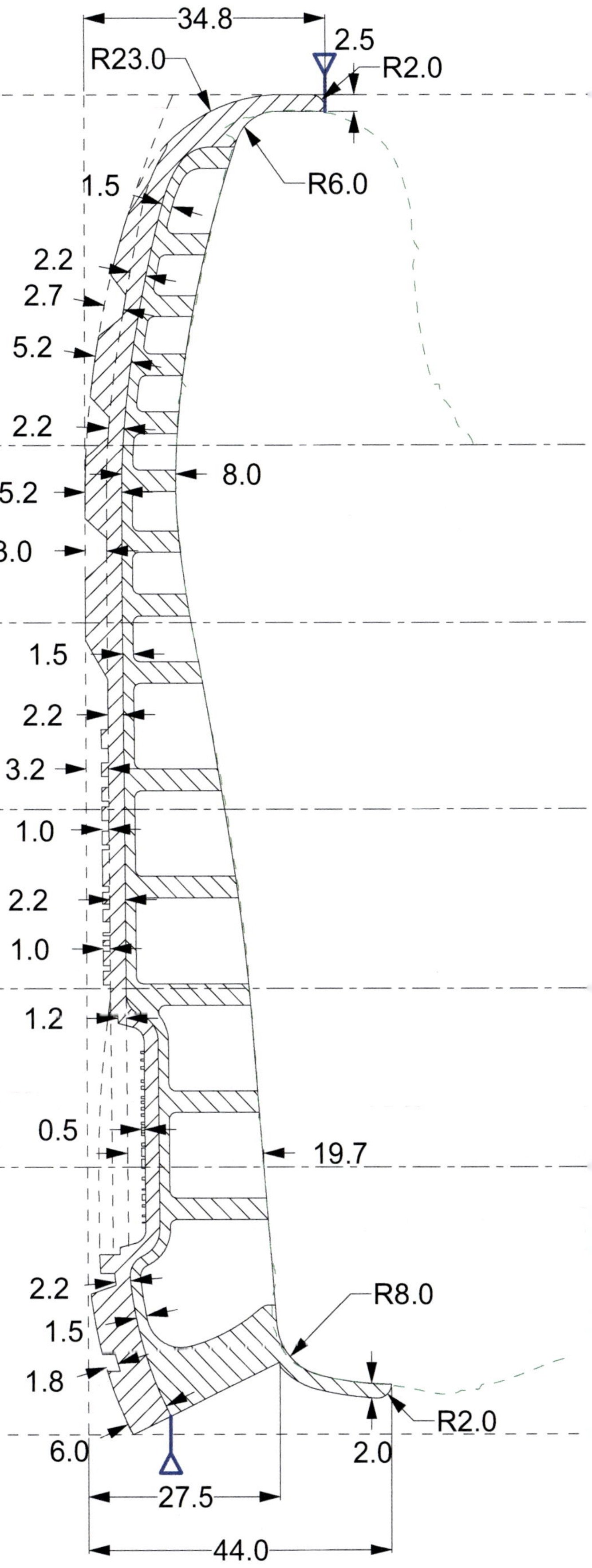

SECTION A-A'

剖面图

Injection Molding: PVC and TPR Outsole Injection

This section focuses on injection molding, specifically the injection of PVC outsoles.

Now that the principles of injection molding are understood, it is important to explore the various types of molds used in the process.

PVC Outsole Injection

PVC is a widely used outsole material, particularly for inexpensive footwear. While it is not ideal for high-performance athletic shoes due to its weight and lower durability, it remains a cost-effective option for general-purpose footwear. PVC may be the Rodney Dangerfield of our materials, perfectly suitable, but it gets no respect.

PVC Injection Mold Blueprint

An injection mold blueprint follows the same principles as those used for compression molds in EVA or rubber components.

Key considerations include:

Wall Thickness Uniformity: Ensuring even wall thickness throughout the component prevents warping during cooling.

Material Distribution: Avoiding excessively thick sections helps reduce weight and material consumption.

A common technique to reduce material use and weight is the inclusion of hollow sections within the outsole. These cavities, often resembling an "egg crate" structure, do not provide cushioning but instead reduce unnecessary material usage while maintaining structural integrity. The solid perimeter of the outsole is designed to enhance bonding.

Dual-Color PVC Injection Machines
A dual-color PVC injection machine allows for the production of multi-colored outsoles.

The process involves:
Injection of the first color.
Rotation of the mold.
Injection of the second color.

Unlike high-pressure plastic injection molding, PVC injection requires significantly lower pressure, making the mold construction less demanding.

Mold Characteristics and Adjustments
Injection molds feature designated channels, known as runners, through which plastic flows into the mold cavity.

During the initial testing phase, adjustments may be necessary to ensure optimal material flow. This can involve widening channels to enhance distribution or incorporating air vents to prevent trapping air within the mold.

TPR Outsole Injection

TPR Injection Molding
Thermoplastic Rubber (TPR) is another material used in footwear manufacturing, similar to PVC but with different processing requirements.

The primary differences between TPR and PVC molding includes:

Machinery Differences: TPR injection machines typically use rotary tables and vertical injection heads.

Lower Injection Pressure: The force required to inject TPR is lower than that of PVC, making the process more cost-effective.
TPR molds are generally inexpensive with costs ranging between $800 and $1,500 per mold.
The production process often involves two-color injection, achieved through a butterfly-style mold that allows for sequential injections of different colors.

While TPR is slightly lighter than PVC, it remains a relatively heavy material compared to EVA or foam alternatives. It is commonly used for budget-friendly footwear, including casual shoes and sandals.

$800 to $1500
per size

Two color butterfly mold

Chapter 15

PVC Outsole Direct Injection

PVC direct injection is a great way to make decent, practical, and cost-effective shoes...... but not **great** shoes.

PVC sole
Aqua sock
$800 to $1500
per size

$5000
per size
1096
SX

L-ZEUS N-ESC
42D
L ZEUS
N ESCOLAR
42-1
42
1
2005

PVC Direct Injection
PVC direct injection is a manufacturing technique where the shoe upper is placed onto a last and the outsole is injected directly onto it. This eliminates the need for separate outsole production and adhesive bonding, improving efficiency.

Key Advantages:
Well-suited for cost-effective footwear such as aqua socks and canvas shoes.
Mold materials are typically steel for durability.
Utilized in carousel machines, where shoes rotate through multiple injection stations for continuous production.

A dual-color direct injection process enables the production of outsoles with multiple colors while maintaining a seamless bond between the layers. Machines performing this process operate with high productivity, producing inexpensive shoes at scale.

Injection molding remains a fundamental technique in footwear manufacturing with PVC and TPR injection processes offering efficient and cost-effective solutions for budget-conscious footwear. While these materials lack the lightweight and performance attributes of EVA or TPU, they provide durability and affordability for mass-market shoes.

The evolution of dual-color injection and direct outsole molding enhances production efficiency, reducing material waste and assembly time. These methods are widely used in markets requiring economical footwear solutions, demonstrating the adaptability and efficiency of modern injection molding technologies.

Chapter 16

PU Foam Outsole Direct Injection

The same direct attach idea but using modern materials and a million dollar machine.

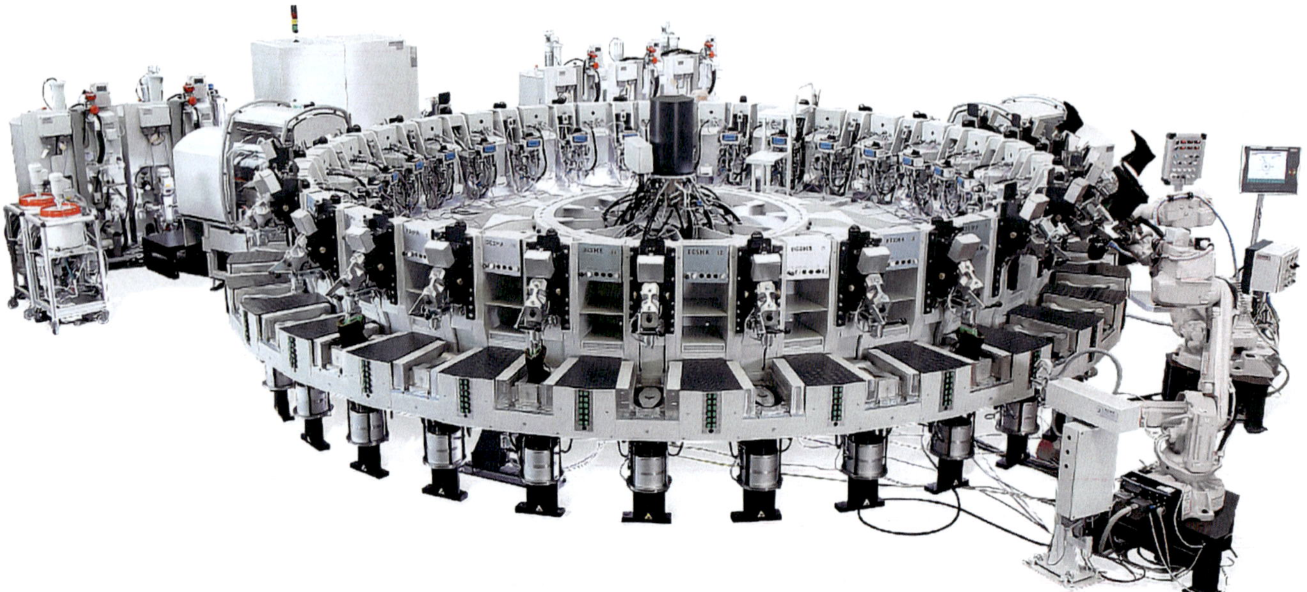

Dual Density PU Injection
We are now moving up the price scale. Previously, we discussed PVC direct injection, and now we will explore the other end of the spectrum—polyurethane (PU) foam outsole direct injection.

The earlier machines we examined were simple carousel machines, appearing somewhat outdated.

In contrast, the machine used for PU injection is a high-end, million-dollar piece of equipment. This particular machine, manufactured by Dema, features dual compound injection capability, allowing it to inject both a bottom rubber-like compound and a foam midsole compound.

This machine has 26 stations, meaning it can produce a pair of shoes in 13 different sizes simultaneously, making it highly efficient for production. Additionally, it incorporates a robotic system that roughs the upper to improve adhesion between the PU and the shoe upper.

PU Injection Process
The process begins with placing the last onto the machine with the upper attached. A mold is then clamped around it. This mold consists of two side panels and a bottom plate. Opposite the upper, a dummy plate is positioned to create the outsole. Once the outsole is formed, the dummy plate is removed, making space for the midsole injection.
This setup ensures that both the outsole and midsole are injected in a single revolution around the machine, significantly improving efficiency.

Machine and Mold Setup
The last is secured using a mechanical actuator. By turning a handle, the last is tightened onto the shoe. Once the shoe is in place, it is crucial to align everything correctly, as this is the only opportunity to ensure precision before injection begins.

The injection process produces a raw outsole which initially has excess material (flashing) that needs to be trimmed. This is similar to rubber pressing, where excess material is removed to refine the final product.

This technology does not come cheap, molds are $6,000 to $10,000 per size.

Midsole mold with clamping edge

Outsole
mold
ZAGO
3732/A
SX 41/1

Automatic edge buffing robot

Almost finished, just needs trimming

ART. ADDAMS
COD: 8014519
TG 33

Chapter 17

PU Outsole Injection

A new high-tech and colorful replacement for rubber outsoles.

PU injection is fast but doesn't have exactly the same character as vulcanized rubber.

The Rise of Transparent PU Outsole Injection Technology
The footwear industry is embracing a new wave of innovation with the introduction of transparent polyurethane (PU) outsole injection molding. This advanced material is reshaping sneaker and performance shoe design, offering durability, flexibility, and aesthetic appeal in one seamless solution.

Unlike traditional rubber or opaque PU outsoles, transparent PU delivers crystal-clear visuals, making it an ideal choice for high-end sneakers, athletic shoes, and fashion-forward designs. This breakthrough material allows brands to create sleek, futuristic styles, with customizable colors, embedded graphics, and even glow-in-the-dark effects, without compromising performance.

From a functional standpoint, injection-molded transparent PU offers superior abrasion resistance, shock absorption, and grip.

Leading manufacturers are also exploring bio-based PU alternatives, reducing environmental impact while maintaining top-tier performance. As demand for both performance and sustainability grows, transparent PU outsoles are becoming a go-to choice for designers seeking cutting-edge style and high-performance durability.

This new material solves two problems with rubber: traditional rubber pressing is slow and the the compound is heavy.

Yes, these injection molds cost about twice the price of a compression mold, but their productivity is more than twice.

$2500
per size

Chapter 18

EVA Injection Molds

A very practical manufacturing process. Oh yeah, you can also make shoes out of it.

TL-12
M13

EVA Injection Molding Process
EVA injection involves injecting material into a mold, where it undergoes a controlled expansion process. The molds are securely bolted to the machine to accommodate the rapid expansion of the material.

Material Injection:
The mold, initially compact, receives the EVA compound, which contains a blowing agent. The mold is kept closed while heat activates the blowing agent, initiating controlled expansion. Blowing agent is simply a chemical compound that when heated breaks down into a bubble of carbon dioxide. This process is what transforms EVA plastic into EVA foam

Rapid Mold Opening:
Once the EVA material reaches its critical expansion state, the air bubble has been created, but the mold pressure is holding it as a tiny bubble. The mold opens rapidly.
The part expands instantly, similar to popcorn, achieving its final size and structure. The mold must open incredibly fast so the EVA part is not damaged by hitting the mold when it expands. EVA injection parts will even have special features designed into the mold to help pull the EVA out so that it can expand safely.

Expansion Ratio Considerations:
The initial cavity size is significantly smaller than the final product due to expansion. When you look at an EVA injection mold, all of the features are half size which can make texture and perforations complicated.

This process enables the creation of lightweight, cushioned components, like midsoles and full shoes.

In this example mold, you can see the springs used to help drive the mold apart.

The mold inserts that form the inside of the shoe are also spring loaded so that as the mold opens, they moved down quickly, allowing the upper to expand off the tooling without damage.

EVA Injection Molding Equipment
The EVA injection machine operates using multiple stations, typically arranged in a press bank. Machines can range from two stations to larger configurations with 12 to 20 stations, increasing production efficiency.

Mold Costs:
Molds can range from $3,000 to $6,000 or more depending on complexity and the number of cavities.

Cooling and Shaping:
Immediately after molding, the parts are soft and pliable.

They must be cooled and sometimes placed in fixtures or on lasts to maintain their shape.

Mold Design and Components
EVA injection molds are commonly made from cast aluminum and include several key elements:
Articulating Form: Determines the shape of the shoe's interior and is often spring-loaded for easy release.

Core and Side Panels: Allow for expansion and removal of the final part.

Injection and Air Release Ports:
Ensure material flows evenly and air escapes during molding.
Molds undergo precise machining and hand-finishing to achieve proper alignment and tolerance, ensuring high-quality production.

Complex and Specialty EVA Molding
Certain EVA designs, such as those with intricate perforations (e.g., Yeezy Foam Runners), require specialized molds.

Two-Color EVA Injection:
Achieved through a hydraulic sliding plate that separates two material injections.
The mold first injects the bottom layer, then shifts and injects the top before final heating and expansion.
This technique allows seamless bonding of two colors.

2 station press bank

Advantages and Limitations of EVA Injection Molding

Advantages:

One-Piece Construction: Eliminates seams and adhesives, improving durability.

Lightweight and Comfortable: EVA is cushioned and flexible.

High Productivity: Once the mold is established, production scales efficiently.

Customizability: Wide range of colors and designs can be achieved.

Limitations:

Expensive Molds: Initial tooling costs are high.
Design Constraints: While versatile, extreme designs may not be feasible.

Shrinkage and Warping: Requires cooling fixtures to maintain shape consistency.

Conclusion

EVA injection molding is a transformative technology in footwear manufacturing, offering durability, efficiency, and customization. While the process involves intricate mold designs and costly tooling, the high productivity and lightweight properties make it a valuable choice for various shoe designs.

With advancements in two-color injection and complex mold configurations, EVA injection continues to evolve, providing innovative solutions for both casual and performance footwear.

Here (photo on left) you can see two mold technicians working on the inner core of an 3D EVA injection mold.

This unique design requires the mold to open in a different direction

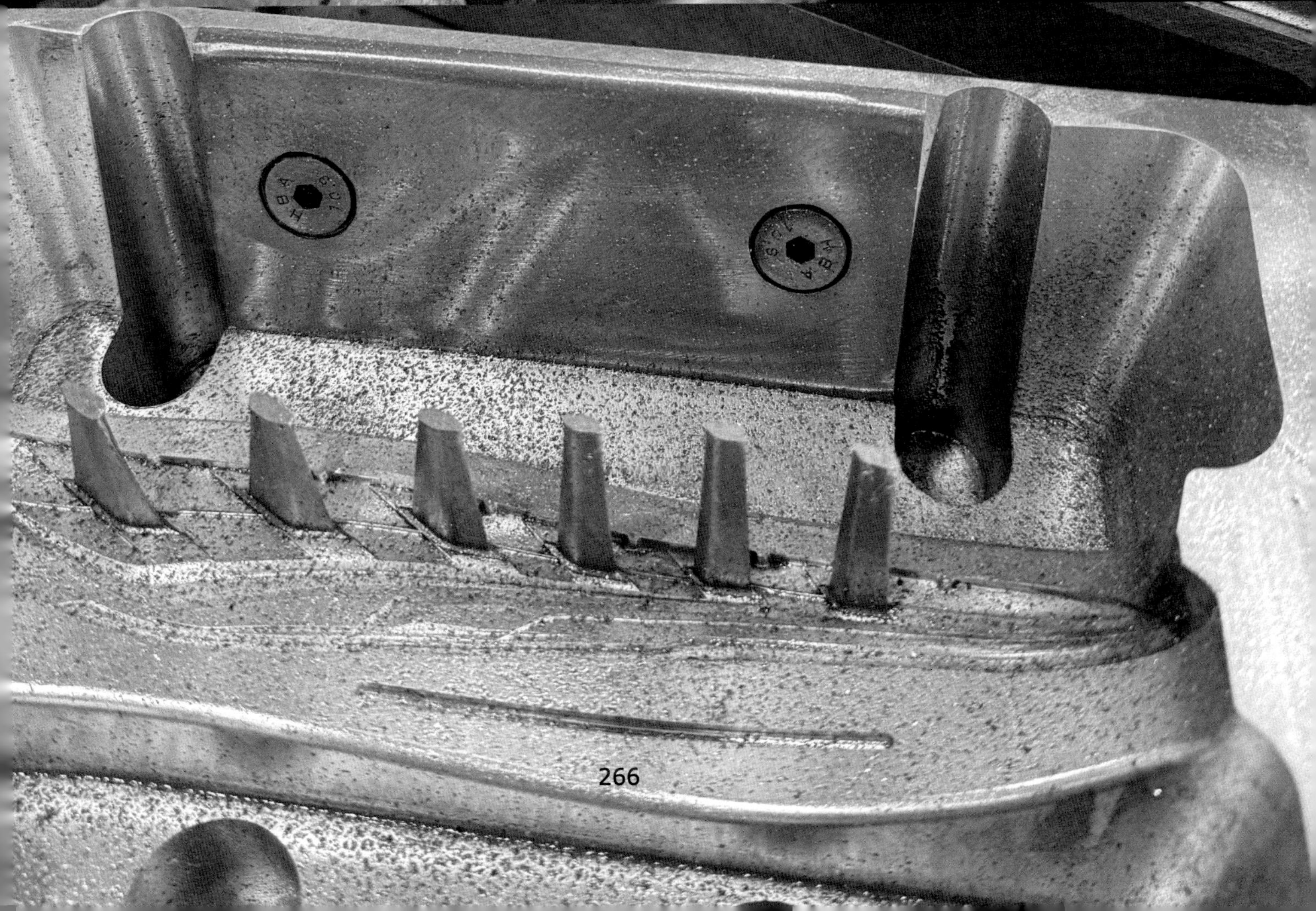

Casey
R
R

DATE

The last is the cooling fixture

Cooling fixtures

Chapter 19

Blow Molding Air Bags

The airbag is a perfect example of how existing technologies can be modified to the requirements of footwear.

Airbag Manufacturing Process in Footwear
Airbags are a unique component in modern footwear, particularly in cushioning systems like Nike Air Max. They are primarily produced through blow molding, a process that shapes plastic into hollow, durable structures.

The Blow Molding Process
The manufacturing process for footwear airbags involves several key steps:

Extrusion of Plastic Tube (Parison Formation)
A plastic tube, known as a parison, is extruded downward from the machine.
The parison remains soft and flexible as it descends.

Mold Clamping and Inflation
The mold clamps around the parison, shaping it into the desired airbag form. A small air tube punctures the plastic and injects pressurized air, forcing the material against the mold walls.

Airbag Formation and Cooling
The airbag maintains its shape as it cools and solidifies inside the mold. Features like dimples or treads are formed by mold impressions.

Trimming and Sealing
Excess plastic from the mold edges is trimmed. Some airbags are pre-sealed, while others are filled and sealed later using an RF (radio frequency) welding process.

Key Components of Airbags
Materials:
Most airbags are made from thermoplastic polyurethane (TPU), offering durability, flexibility, and resistance to degradation.
Some lower-cost versions may be made of polyvinyl chloride (PVC).

Airbag Coloring:
Airbags are naturally transparent and achieve their final color through internal painting.
A spraying fixture is used to apply pigment evenly inside the airbag. Tinted plastic can also be used.

Air Filling and Final Sealing:
A worker inserts a needle into the airbag to introduce air or nitrogen.
The RF welding machine then seals the airbag shut to maintain its structure.

Cooling lines

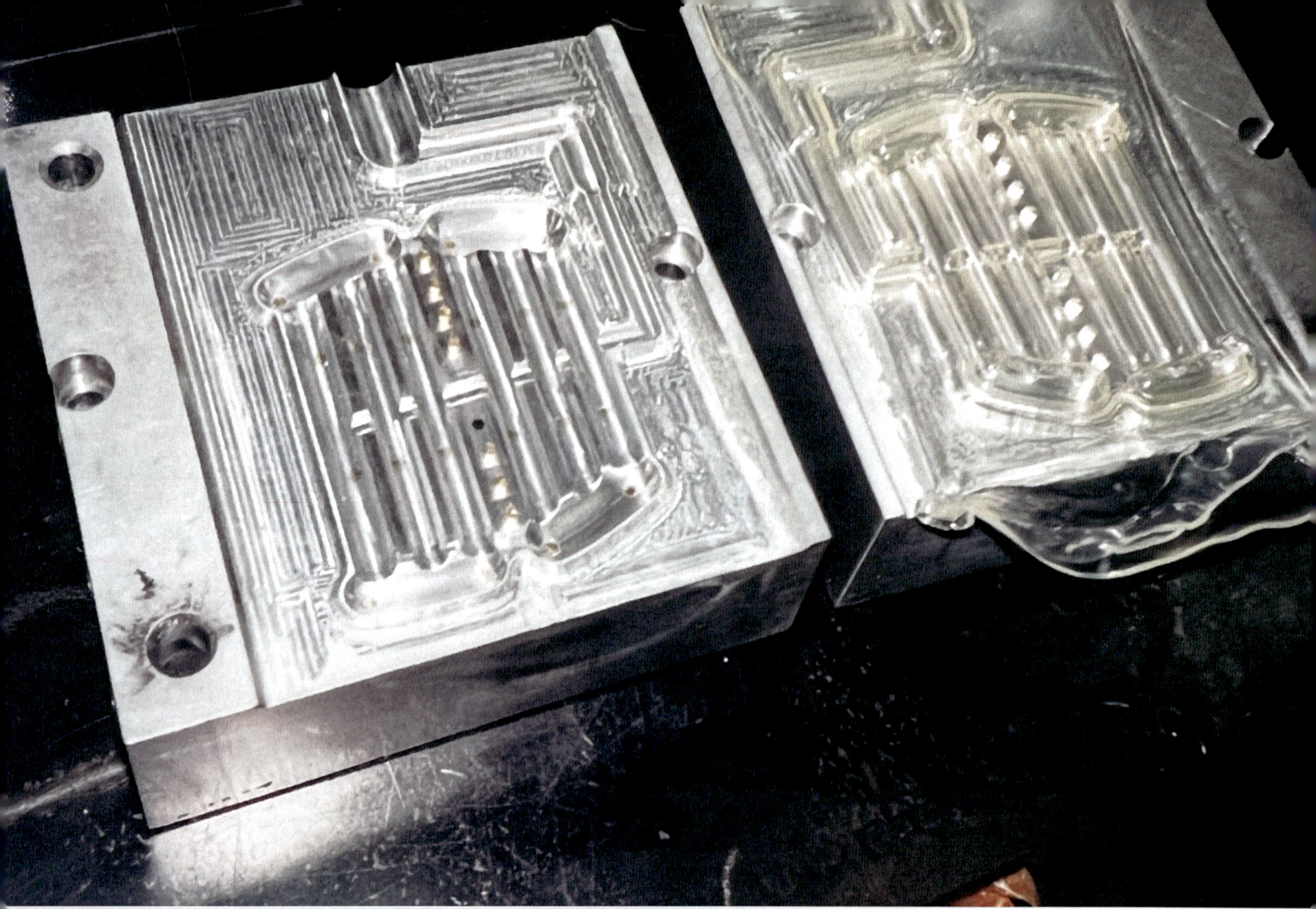

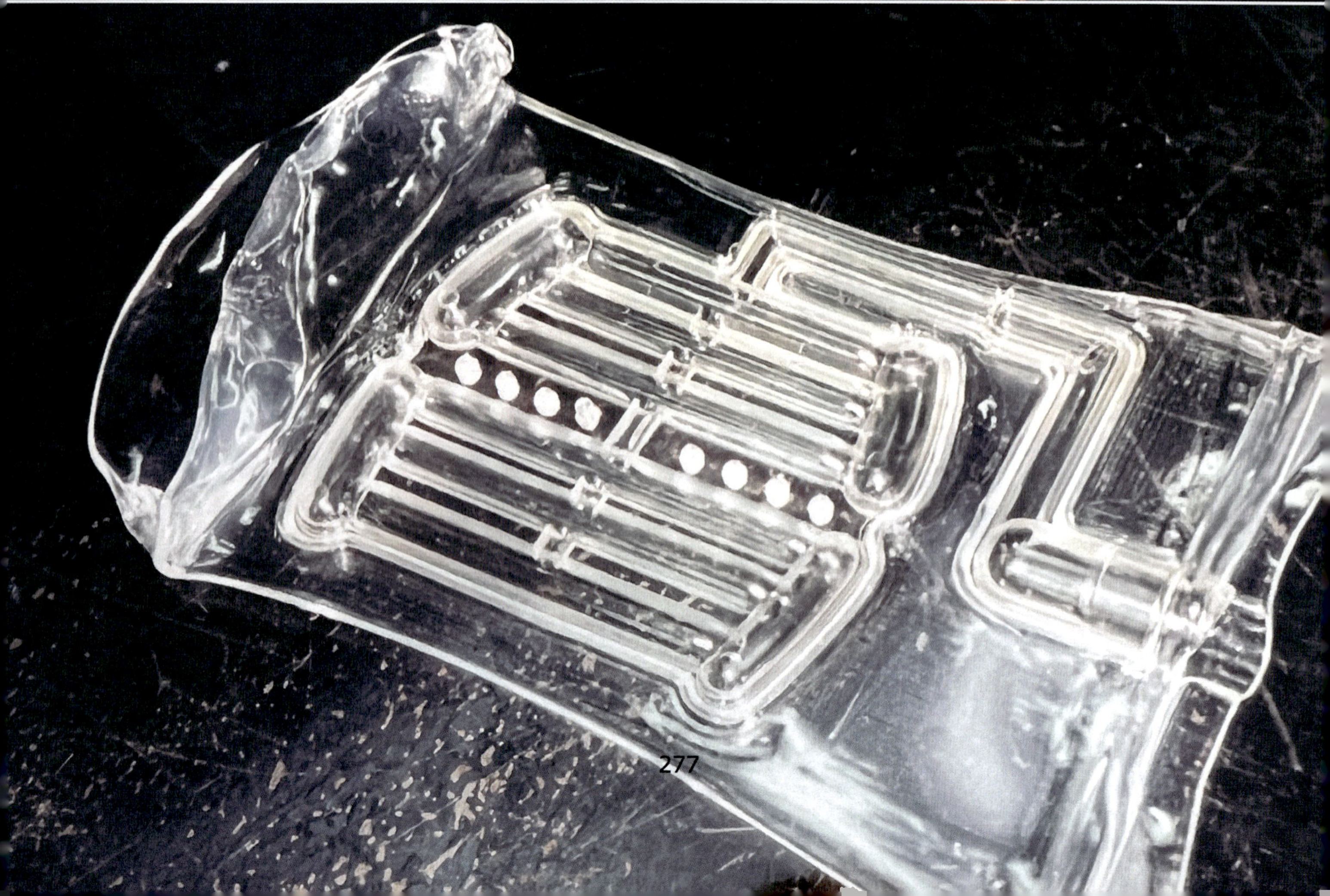

Different Types of Airbags and Their Manufacturing Variations

Basic Airbags (e.g., Nike Air Max units)
Manufactured through blow molding with simple parting lines.
Molded in two halves that are clamped together.

Advanced Asymmetrical Airbags
Feature strategically placed treads and variations in shape for optimized performance.
Require more precise control over parting line placement.

Nike Zoom Airbags
(Alternative Manufacturing Method)
Created by welding fabric inside the airbag, rather than solely relying on blow molding.
A perimeter welding process is used to reinforce the structure before inflation.

Trimming fixture

Filling station

Tooling Costs and Considerations
Mold Costs:
Basic airbag molds typically range between $1,500 and $10,000, depending on complexity.

More intricate molds, with additional features like asymmetrical designs or reinforced treads, can cost significantly more.

Production Precision:
Mold alignment is critical to ensuring proper air retention and structural integrity.
A slight misalignment in the molding process can result in air leakage or improper expansion.

The production of airbags in footwear, while not a new concept, continues to evolve with improved materials and advanced manufacturing techniques. Blow molding remains the dominant process, offering efficiency and scalability for large-scale production.

Innovations such as fabric-reinforced airbags and customized asymmetrical designs demonstrate how footwear companies continue refining air-cushioning technology to enhance performance, durability, and aesthetics.

Despite the upfront costs of mold production, airbags provide long-term benefits in lightweight cushioning and impact absorption, making them a staple in athletic and lifestyle footwear design.

AIR MAX

Chapter 20

Footwear Production Fixtures

Footwear Production Fixtures
Precision and Efficiency in Manufacturing.

The unsung heroes of footwear production
Footwear production relies on a variety of specialized fixtures to ensure consistency, precision, and efficiency. While they may not seem as exciting as advanced injection molding or EVA technology, these tools play a vital role in achieving high-quality footwear production.

This chapter explores various types of fixtures used in modern footwear manufacturing, particularly in stitching, molding, and shaping processes.

Pay special attention to these fixtures when touring a shoe factory for the first time. These are some of the mundane items that are absolutely required for high-end footwear production.

Computerized Stitching Fixtures
Modern footwear manufacturing employs computerized sewing machines to replace traditional hand stitching. These machines utilize custom plastic or metal fixtures to hold materials in place, allowing precise and efficient stitching.

Fixtures for Complex Stitching Patterns
Traditionally, patterns were silk-screened onto fabric, and workers manually followed stitch lines.
Now, automated machines use customized plastic frames that hold the upper components securely during stitching.

These fixtures are unique to each shoe design and must be specially developed for production.

Process of Computerized Stitching
The shoe upper is placed into the fixture which is then secured onto the sewing machine platform.

Once aligned, the machine automatically stitches multiple layers with high precision. This ensures accuracy, consistency, and reduced labor costs while increasing production speed.

Advantages of Computerized Stitching
Computer controlled stitching eliminates human error in production. Reduces production time and increases output. It also creates a cleaner, more professional appearance, for complex stitch patterns.

Computerized sewing dramatically reduces the training required for operators.

Human operators
Human operators are still required for short runs and some complicated upper closing operations. Of course, no fixture can replace the experience, talent, and flexibility of a highly trained stitcher. The best stitchers will be found in the sample room, not on the production assembly line.

Labor requirements
While China is often considered the home of cheap labor, shoe factories in the far east are still working to reduce the labor cost for modern shoes. In fact, many brands are now training their designers to create shoes that can be made predominantly with computer stitching.

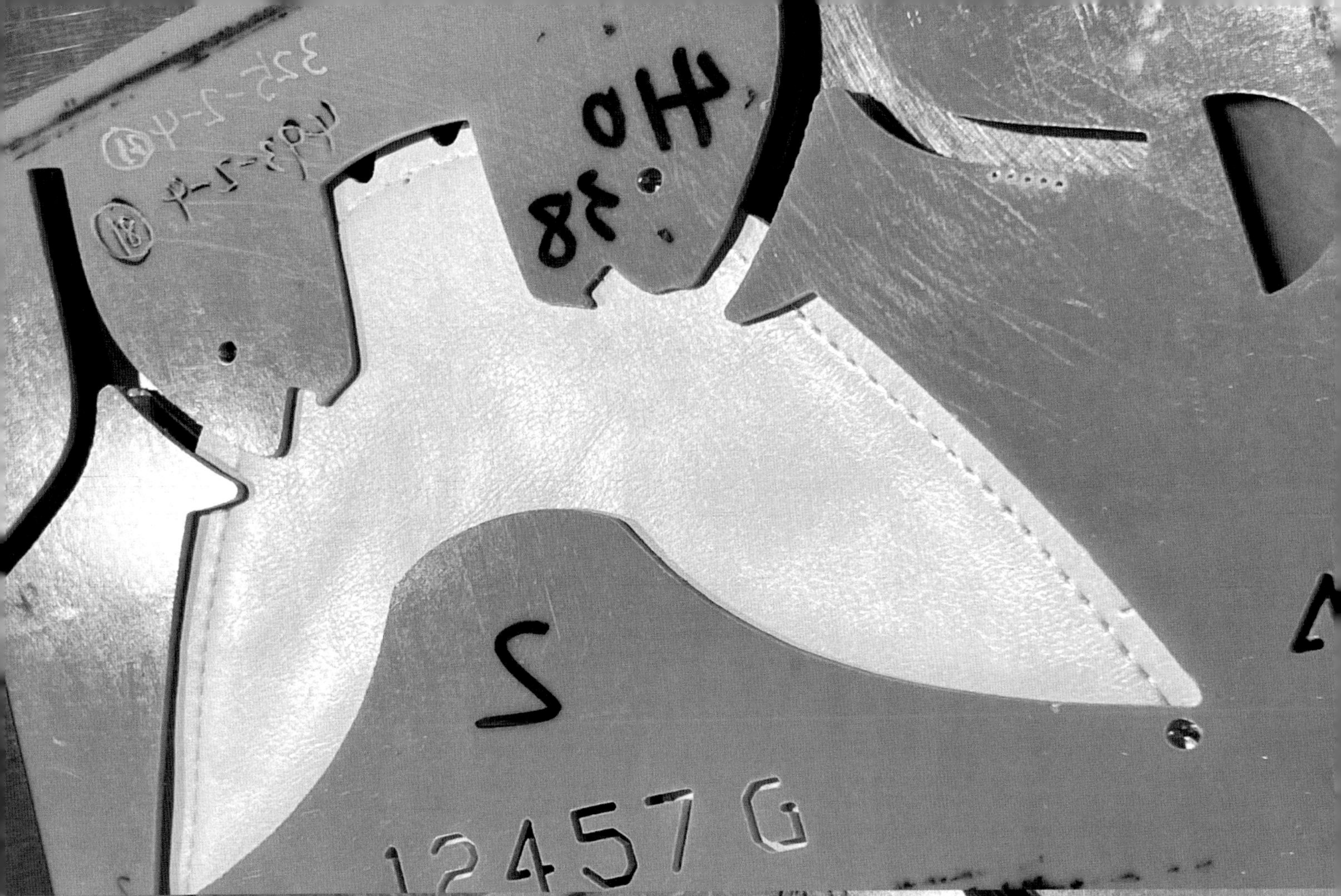
S
12457 G

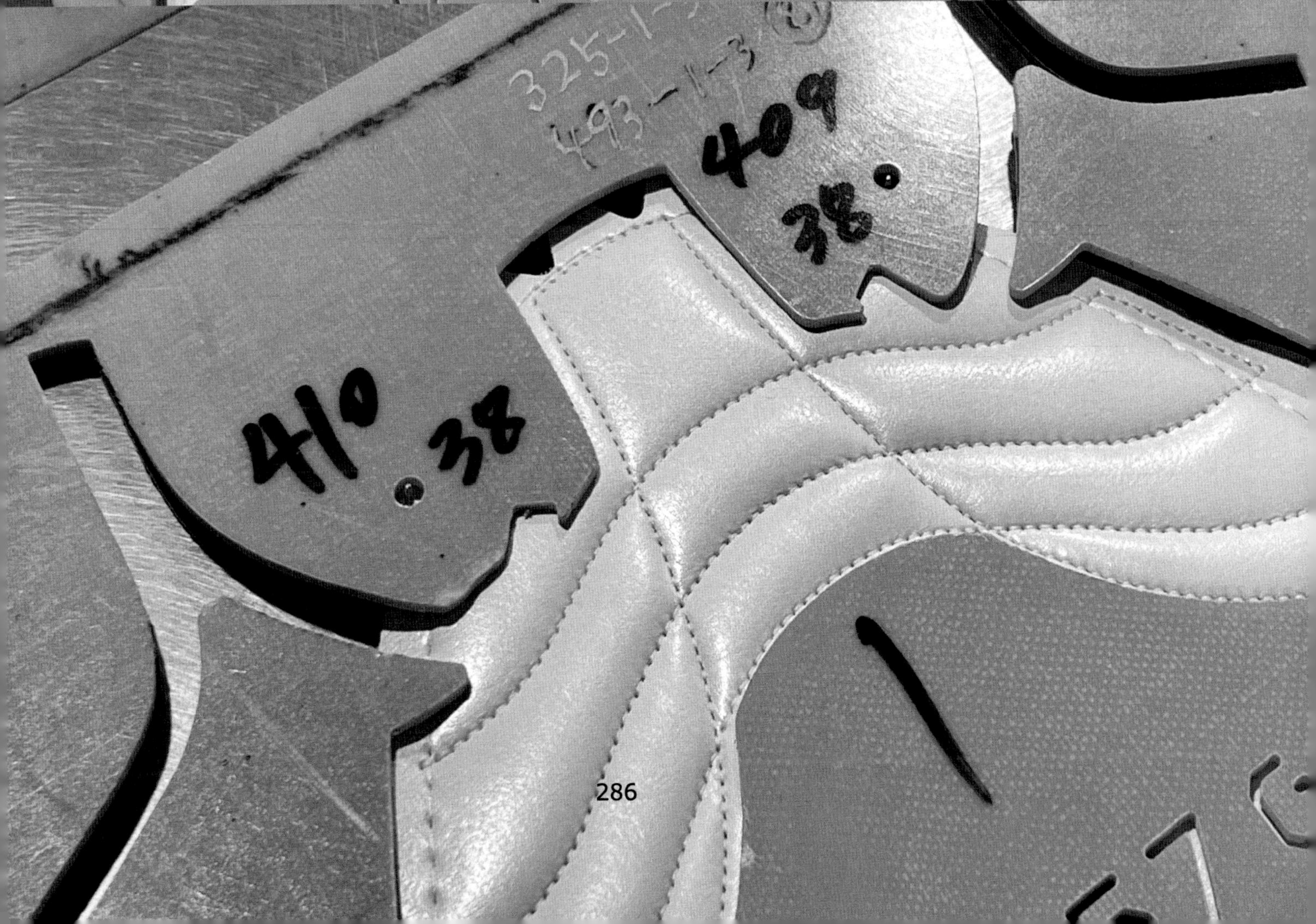
409
38
410
38

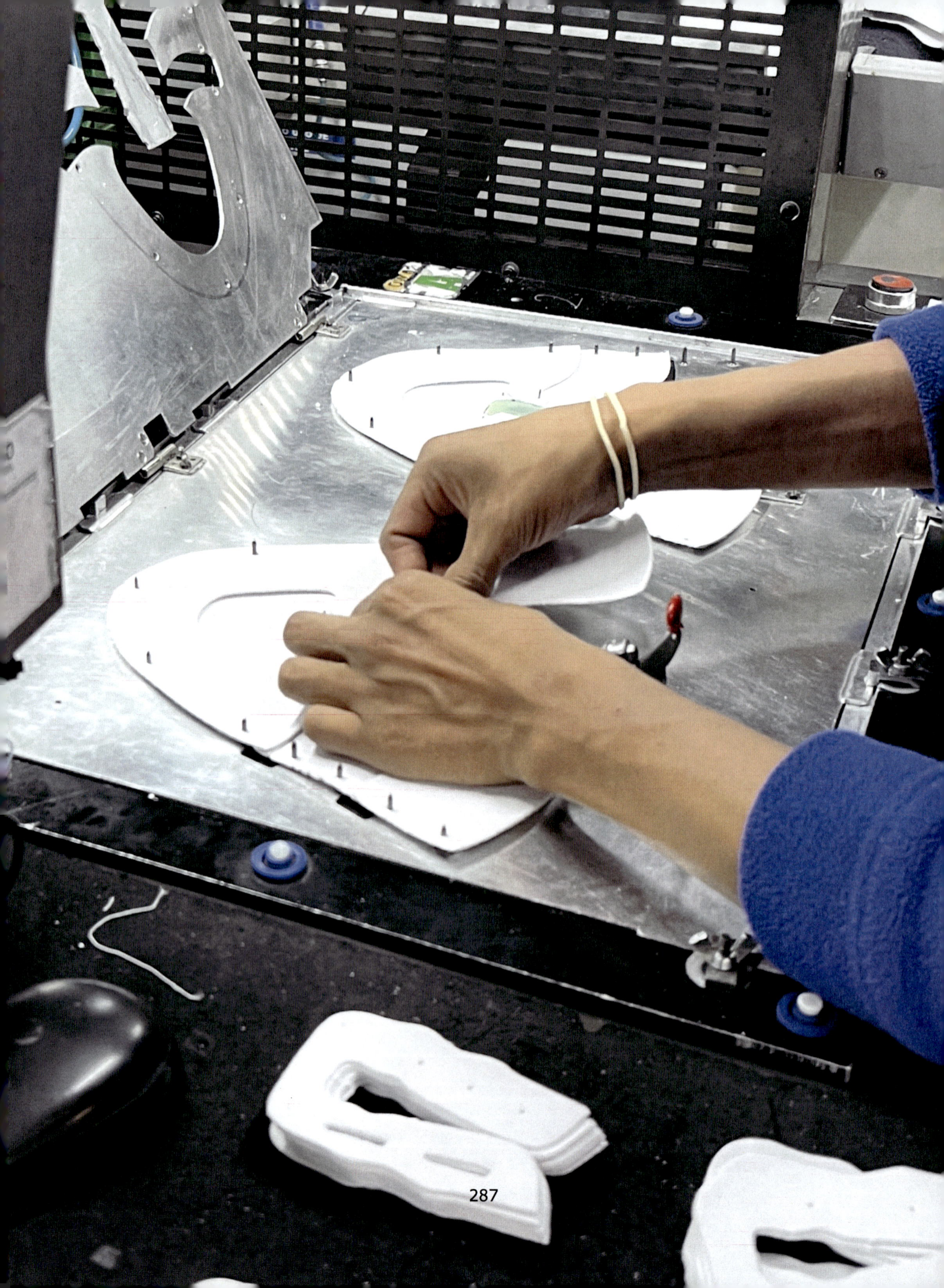

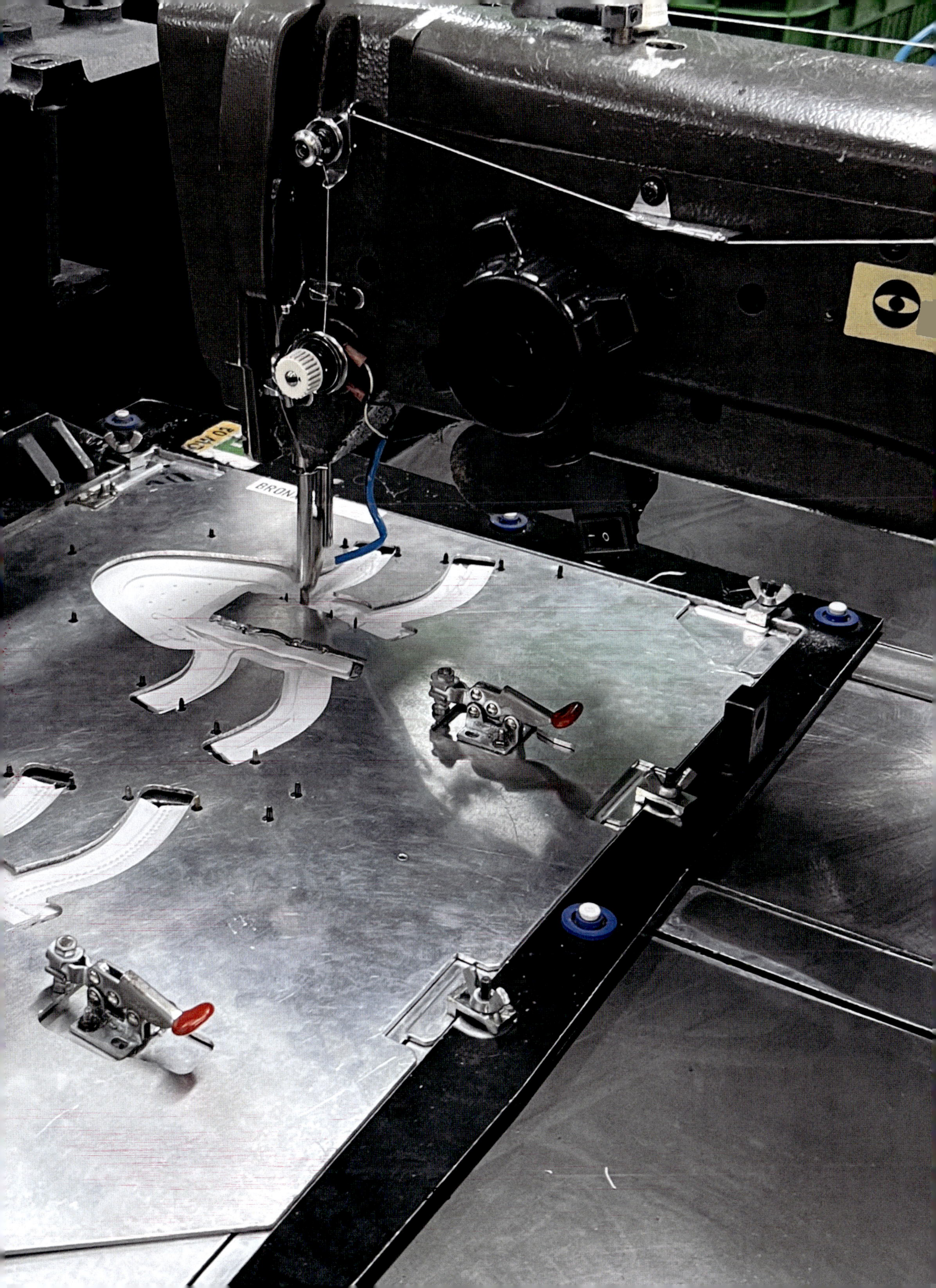

JUKI
(1)

Design and Development team effort
Creating footwear designs for computer stitching processes is a team effort. The shoe designers, pattern developers, and cutting die makers all need to work together to create pin locations, the alignment fixtures, and a process that will hide the perforations that anchor the upper into the machine.

This process is required to manufacture shoes in high labor cost countries. And when I say high labor cost countries, I mean every country in the world.

As your production volume increases, you will definitely need to invest in some sort of automated stitching process.

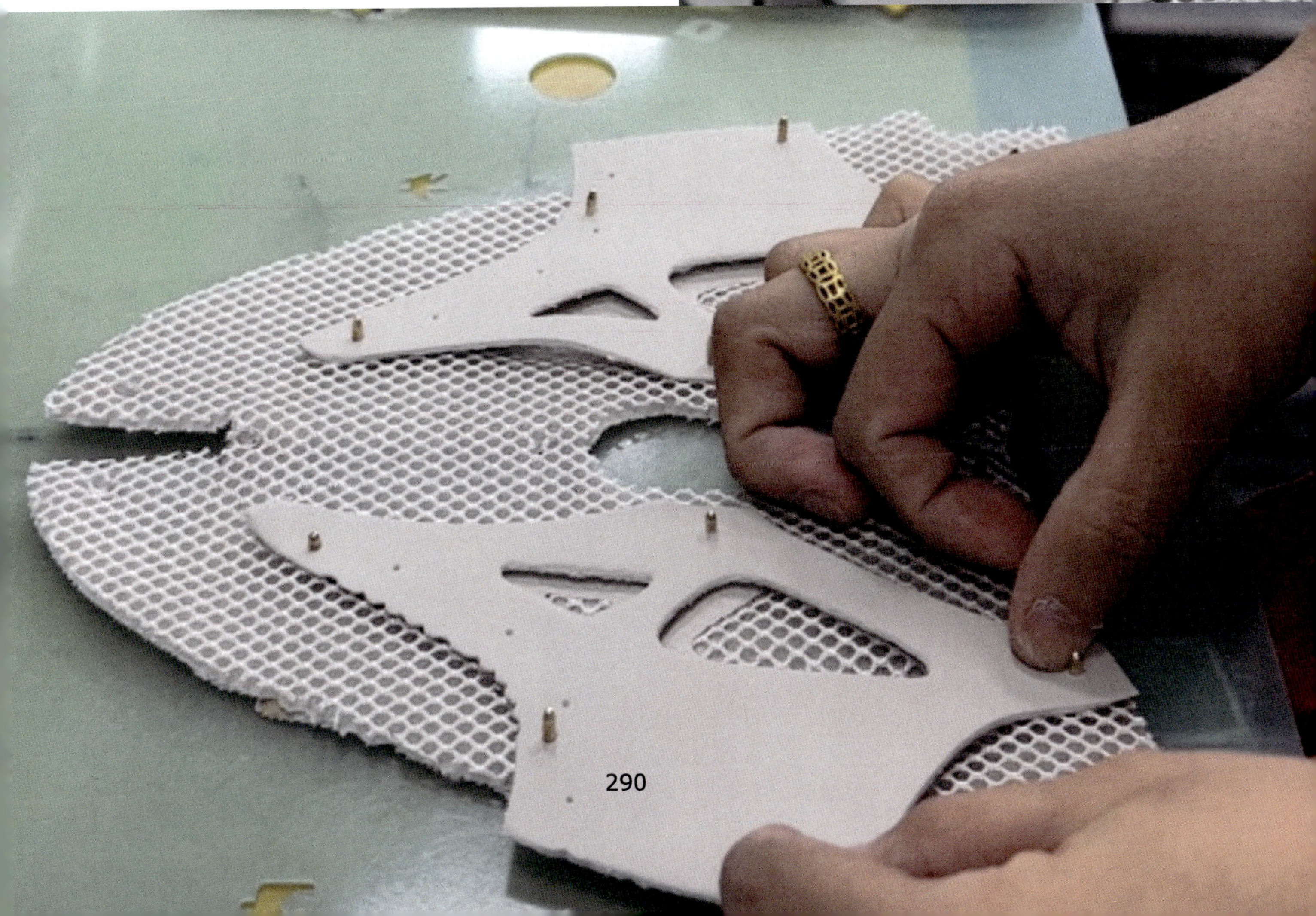

Knit Upper Setting Fixtures

Knitted uppers often lack structural integrity necessary during stitching assembly. When the knit uppers come off the machine, they are a floppy mess.

Workers use template-based fixtures to steam and press the uppers into their correct shape before final assembly.

This process ensures that the knit material remains aligned, structured, and ready for laser trimming and stitching.

Chilled fixture with laser guides

Toe Box Shaping Fixtures
Toe box shaping molds are used to maintain the precise contours of the shoe's forefoot and tip.

These molds ensure that shoes replicate the original last shape consistently across all sizes.

These machines are generally located at the head of the assembly line. The machine features a heater to soften the material, moldable toe reinforcement, and a shape setting fixture that is exactly the shape of the last.

This type of molding machine is absolutely critical for Strobel shoes. Board lasted or mechanically lasted shoes do not require this type of equipment to shape the toe tip.

These fixtures can save you money. It's possible to avoid more expensive board lasting operations if you can shape the toe tip.

Before and after

The heel setter with hot and cold stations

Heel Molding Fixtures
There are many different types of heel setting molding machines and fixtures. You should see a station where the upper is heated and a clamping station with a chilled cooling fixture to set the heel counter shape.

You'll often see heating wires leading to the hot station and frost on the cooling station.

The fixtures should be your last with a certain size run listed.

This type of machine is used to set pre-molded heel counters, but also works perfectly well on sheet cut plastic thermo plastic heel counters.

This type of machine is almost indispensable for modern athletic footwear production. You should see the heel station and the toe forming station at the head of the assembly line.

Because the upper must be open, these operations are done before the Strobel sock is attached.

Again, the heel setting machine is absolutely critical to create a well formed shoe. You will see this machine used on both Strobel and board lasted shoes.

COL-34H

Col-

Sole Pressing Fixtures and Bonding

Once the shoe components are assembled, pressing fixtures ensure strong adhesion between the outsole and midsole without distorting the shoe's structure.

Role of Sole Pressing Fixtures

During bonding, the pressing machine applies force to attach the outsole to the midsole and upper.

Fixtures ensure that pressure is distributed evenly, preventing sidewall dents or gluing misalignment.

Materials and Construction

Sole pressing fixtures are typically made from cast urethane to withstand repeated use without damaging the shoe.

Automation and the Future of Footwear Production

The use of robotic sewing, automated shaping, and precise fixtures is becoming more common, particularly in large-scale production environments.

Increased Productivity:

Automated stitching has significantly reduced the reliance on manual labor, particularly in large factories in China, India, and Bangladesh.

Some manufacturers have transitioned to fully automated stitching operations, resulting in shoes that are flawlessly assembled with fewer defects.

Cost Efficiency:

Labor costs are rising worldwide, pushing manufacturers to replace manual stitching with robotic automation.
While these fixtures require an initial investment, they ultimately reduce costs by improving efficiency and reducing waste.

Fixtures play a critical role in ensuring the consistency, precision, and efficiency of modern footwear production. Whether in stitching, molding, shaping, or bonding, these tools are essential for creating high-quality footwear at scale. With advancements in automation and robotics, manufacturers continue to refine these processes, improving productivity while maintaining superior craftsmanship.

Chapter 21

Radio Frequency Welding Tools

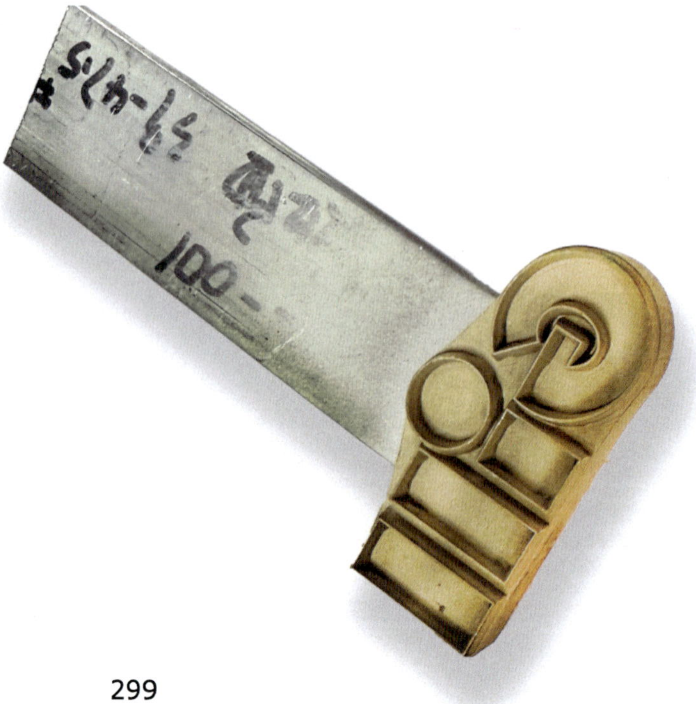

RF Welding (Radio Frequency Welding) in Footwear

RF welding is a technique that uses high-frequency electromagnetic waves to bond synthetic materials like TPU, PVC, and EVA. It is commonly used for creating logos, grip elements, and structural reinforcements.

How RF Welding Works

A brass or copper die is mounted onto an RF welding machine.

When activated, radio waves create heat through molecular vibration within the materials.

The terms "high frequency welding" and "radio frequency welding" are essentially the same thing. They are often used interchangeably, with "radio frequency welding" being the more common phrase.

The process creates embossed logos, reinforcements, and seamless bonds in footwear components.

Applications in Footwear

Logos and branding:
Many 3D logos on synthetic shoes are made using RF welding. This works great on man-made materials that melt. Materials like PVC and PU respond well with crisp embosses and deep relief. Natural leathers respond by branding, natural fibers, such as cotton, do not respond to RF welding.

Grip elements:
Used to add textured grip patterns to outsoles or insoles.

Seamless bonding:
Common in waterproof or seamless footwear designs. Materials such as TPU that bond to themselves are great candidates for radio frequency welding. Nike ZOOM airbags are created by radio frequency welding process.

Advantages of RF Welding

Produces precise, clean embossing without the need for additional adhesives.

Cost-effective and durable, with each brass die costing only a few hundred dollars.

Ideal for mass production of synthetic and athletic footwear.

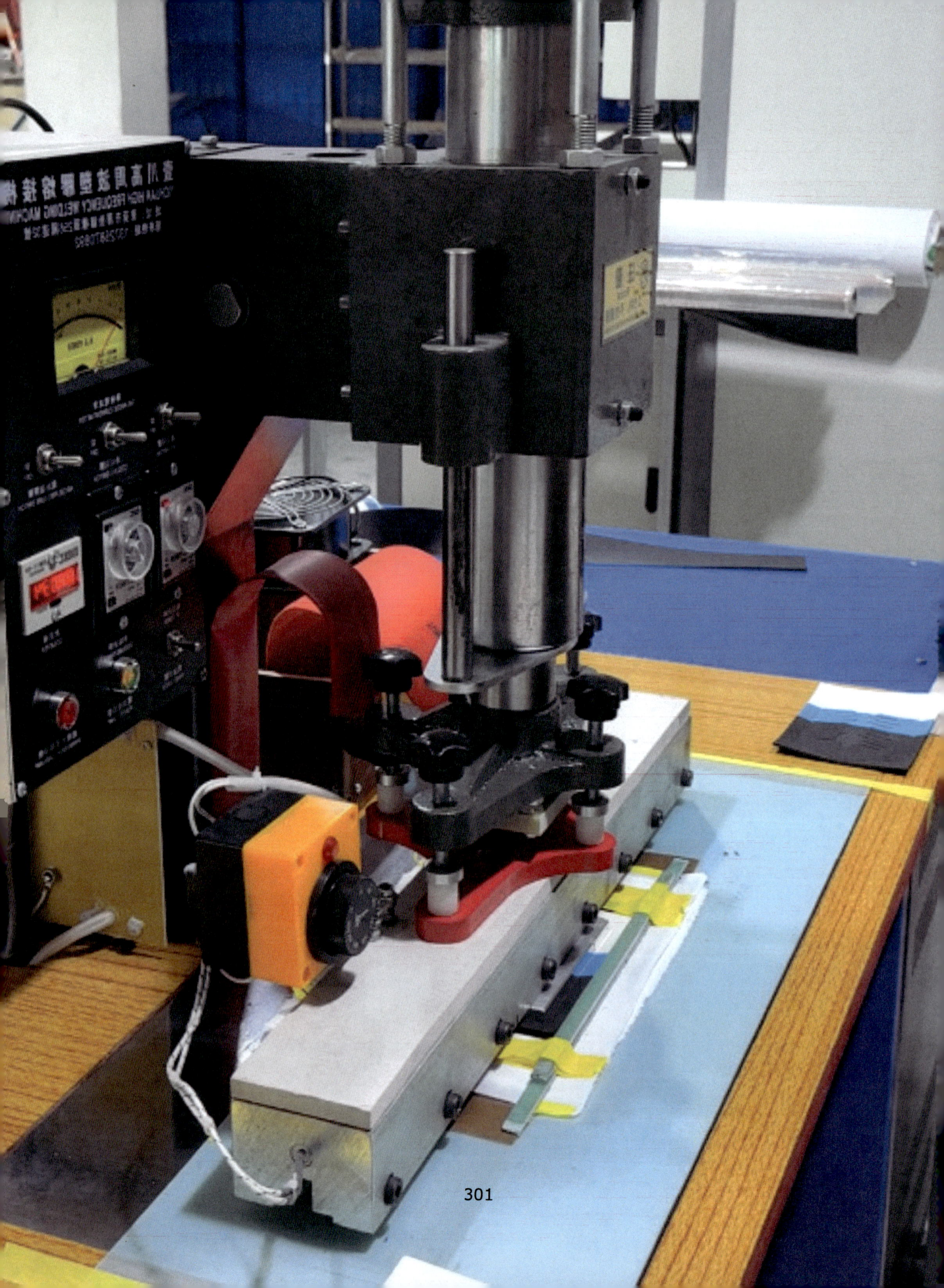

Radio frequency welding machines
Radio frequency welding machines come in a variety of sizes. The desktop machine shown here is very typical of what you'd find in a shoe factory.

All types of footwear logos, big and small, can be done on this type of machine. The welding tool itself is usually made from brass. The factories use brass as it's quite soft and easy to cut and doesn't rust. These molds are quite inexpensive, only $200 or $300 each. Usually, the shoe factory will group grade any logos so you may only need two or three to cover your entire size run, depending on the size, shape, and location of the logo design.

The radio frequency tool can do a few different things. It can emboss a silkscreen logo, it can also fuse and cut TPU film to create a logo.

You can see the logo example to the right has a sharp edge and this cuts the material as it welds it.

The radio frequency operation can also have backing material behind the logo. On the following page, you'll see an example of this.

You will also see a great example of how radio frequency welding could be used to seal TPU to create an air bladder. You will notice in the Reebok pump example the upper material is also welded together with the bladder.

In this case, they use the RF weld to create a stitch flange and seal the TPU bladder simultaneously.

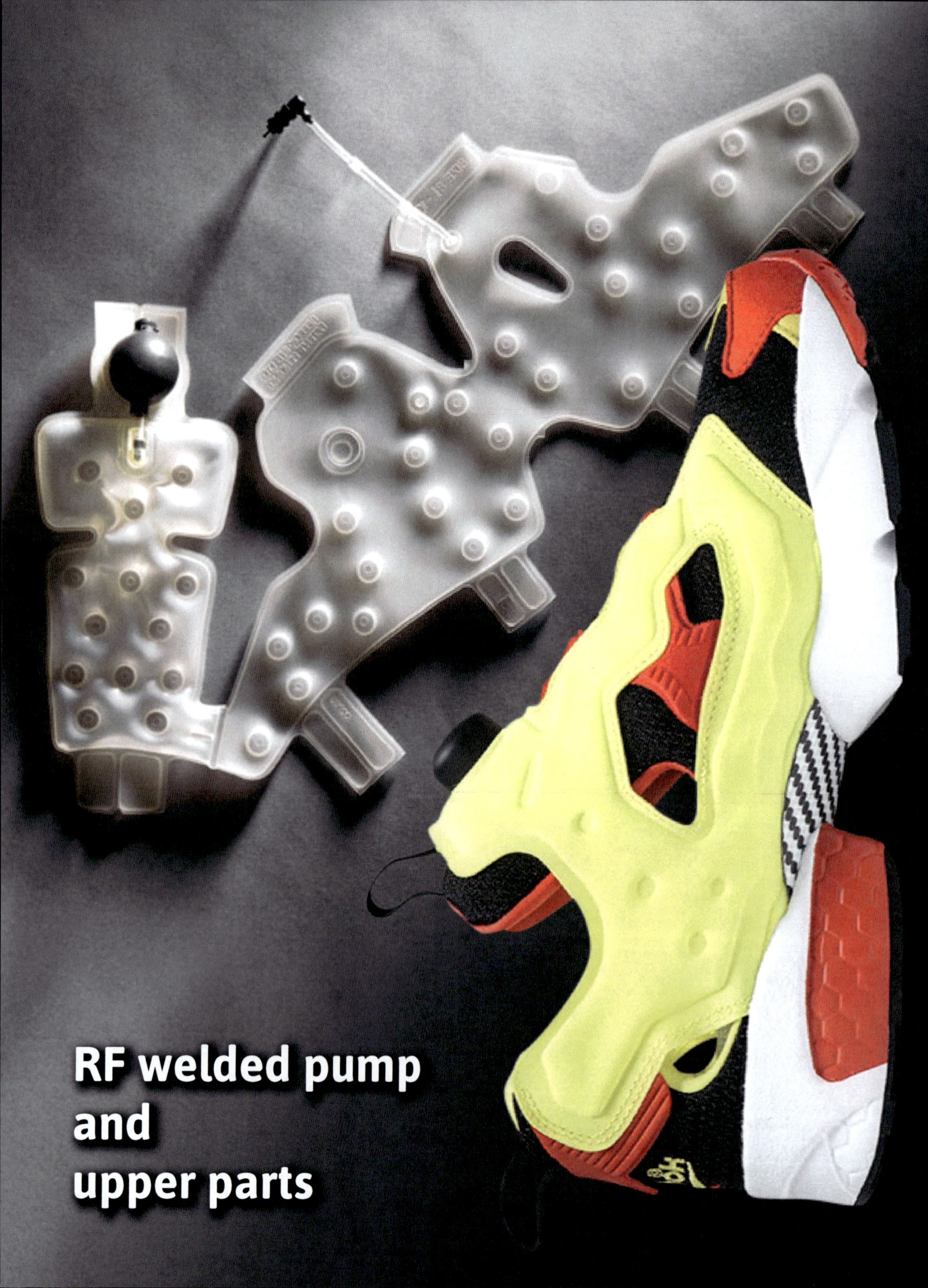

RF welded pump
and
upper parts

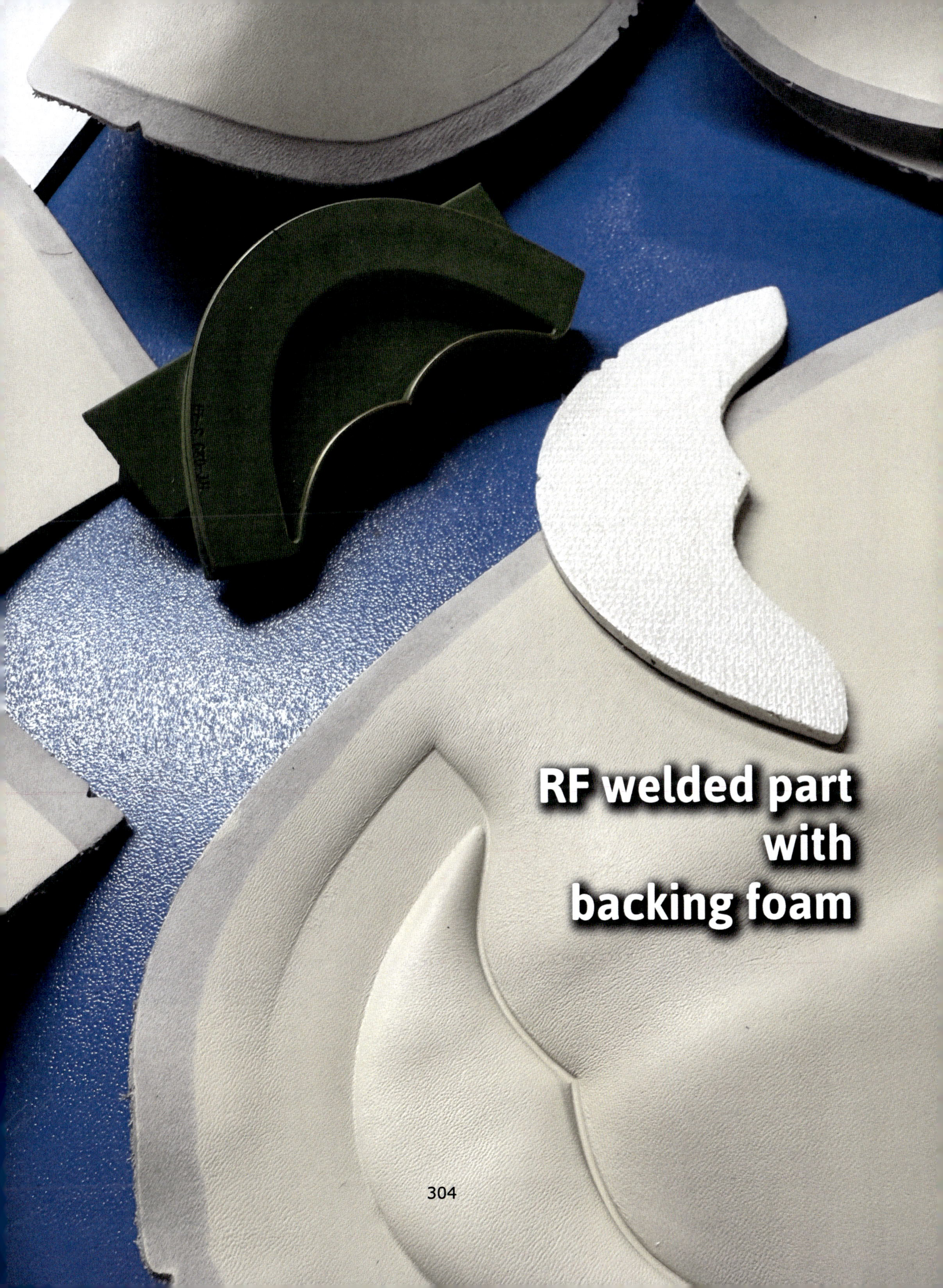
RF welded part
with
backing foam

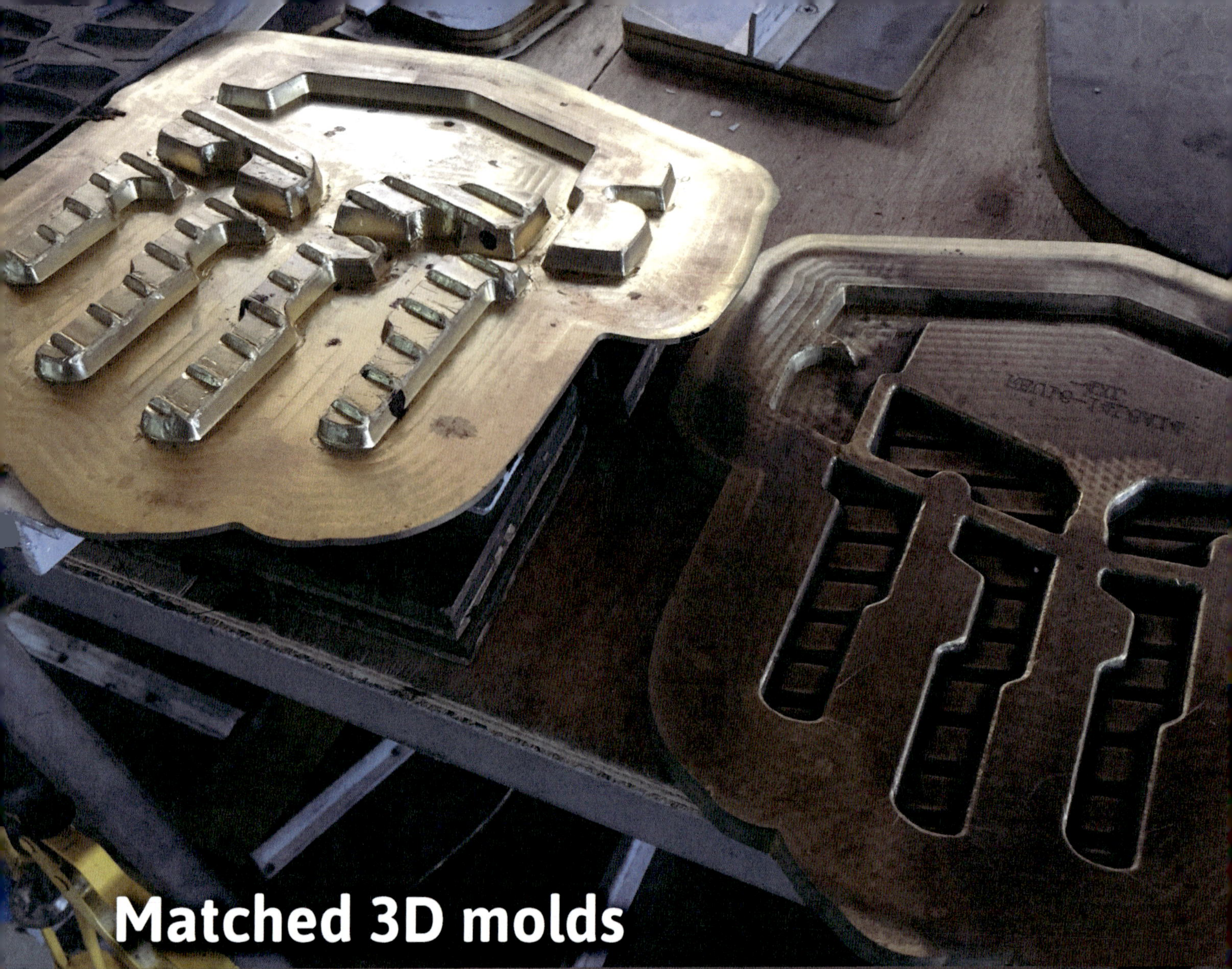

Matched 3D molds

Matched 3D Molds
The radio frequency molding process can also be used to create pockets for gel or other padding materials. To the left, you'll see padding for a sports glove. This set of matched molds is used with a radio frequency machine to create gel pockets.

The example below is a radio frequency mold for an EVA sports grip. The lettering and textures are cut into the brass mold.

Using Radio Frequency Welding
The radio frequency welding process is very common as the equipment, molds, and materials are inexpensive. Not every shoe factory will have welding equipment but there are many subcontractors that do this work.

Welding machines are sized to make small logos and even entire flat shoe patterns.

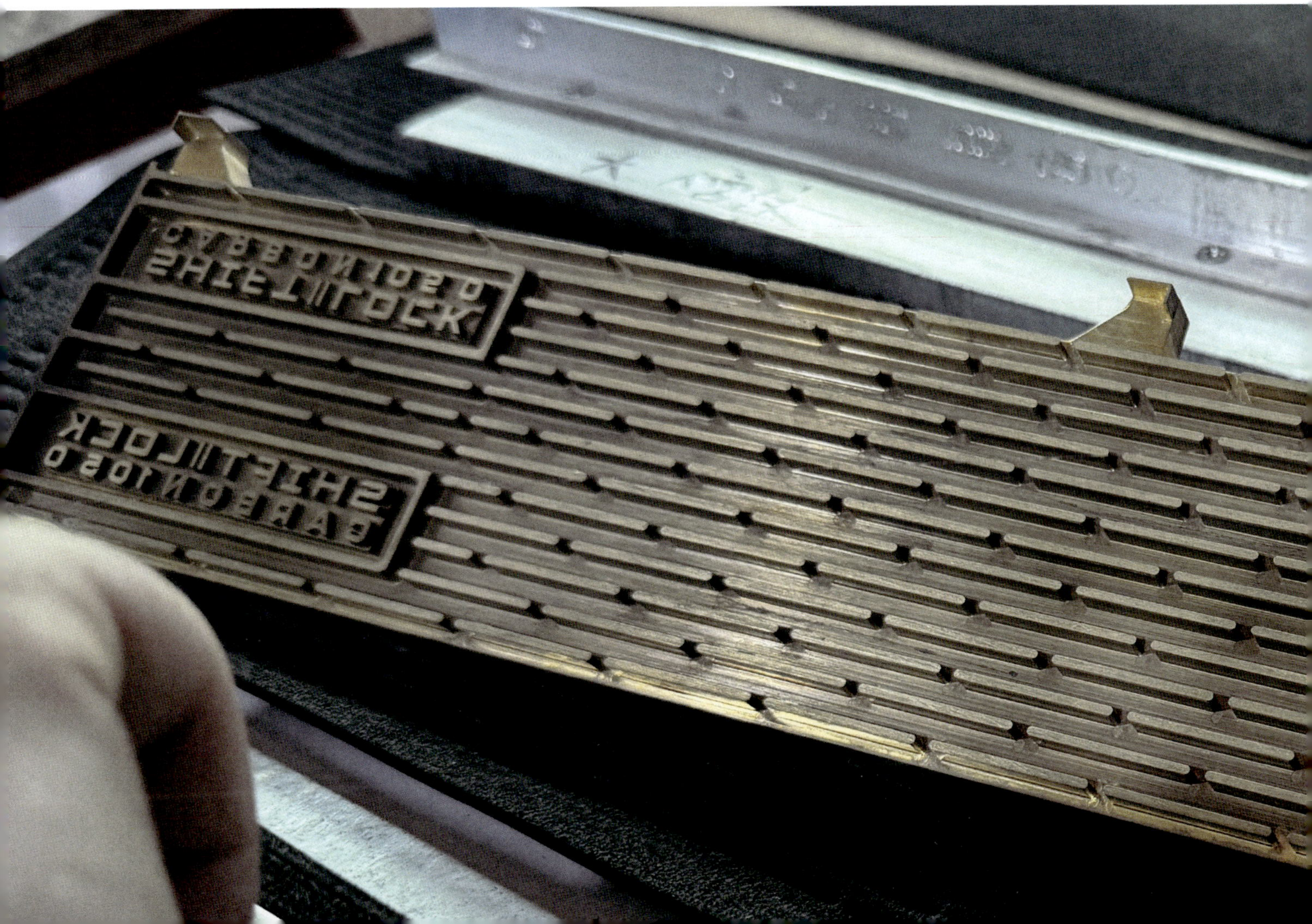

Chapter 22

Hot & Cold Pressing Tools

Heat Pressing in Footwear Manufacturing
Heat pressing is a widely used method for embossing, bonding, and shaping footwear components, particularly for knit uppers and EVA elements.

Heat Pressing for Embossing
Used to create textured patterns and reinforced areas on uppers.

A pressing fixture made of metal or plastic is used to hold materials in place. Workers stack cut materials with fusible backings on the frame, then apply heat and pressure to bond the layers. This process is common in synthetic leather soccer shoes, where the design is screen-printed and fused to the upper.

Shaping EVA Components
EVA (ethylene vinyl acetate) components can be heated in an industrial oven and then pressed into molds to emboss patterns and textures.

A metal pressing fixture shapes the material while cooling, ensuring consistent designs and structural integrity.

Molded Heel Cups and Tongues
Used in high-performance shoes, snowboard boots, and ski boots.

Materials are stacked, pinned onto a mold, and pressed into a three-dimensional shape.
The mold may be heated or cold depending on the process with heated materials setting the final shape upon cooling.

Parts are
assembled and
pressed

Laser cut parts are fused with a heat press

ROAV

Cut parts are assembled flat and pressed into shape

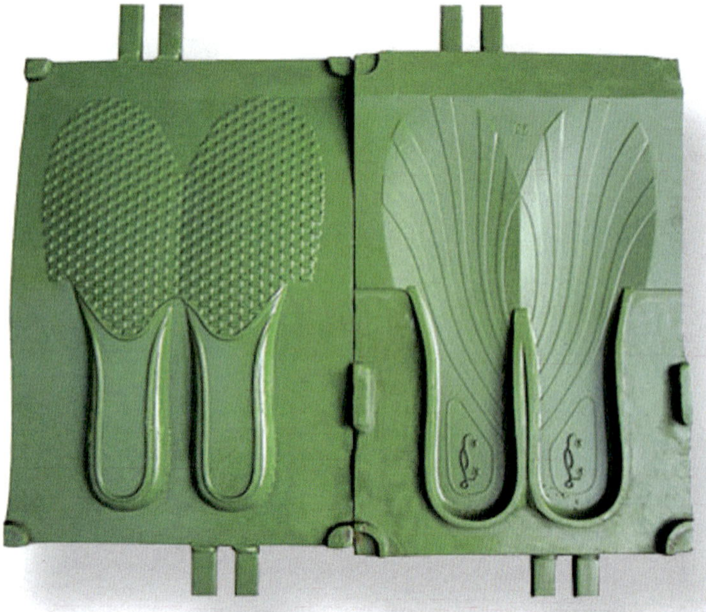

Cold pressed EVA footbeds
This type of cold pressing is very common for EVA footbed production. The EVA components are laminated with fabric, or in this example, memory foam. These parts are heated in what looks like a giant pizza oven.

Once the EVA reaches a critical temperature it is removed from the oven and placed into a cold compression mold. This embosses the EVA and sets the contours of the footbed.

These molds are inexpensive. They cost less than $500 per size. Because the front of this mold is flat, one EVA compression tool can be shared over several sizes of shoes.

You'll notice these molds are bolted to the press. Each mold has alignment features to ensure the top and the bottom are matched when pressing.

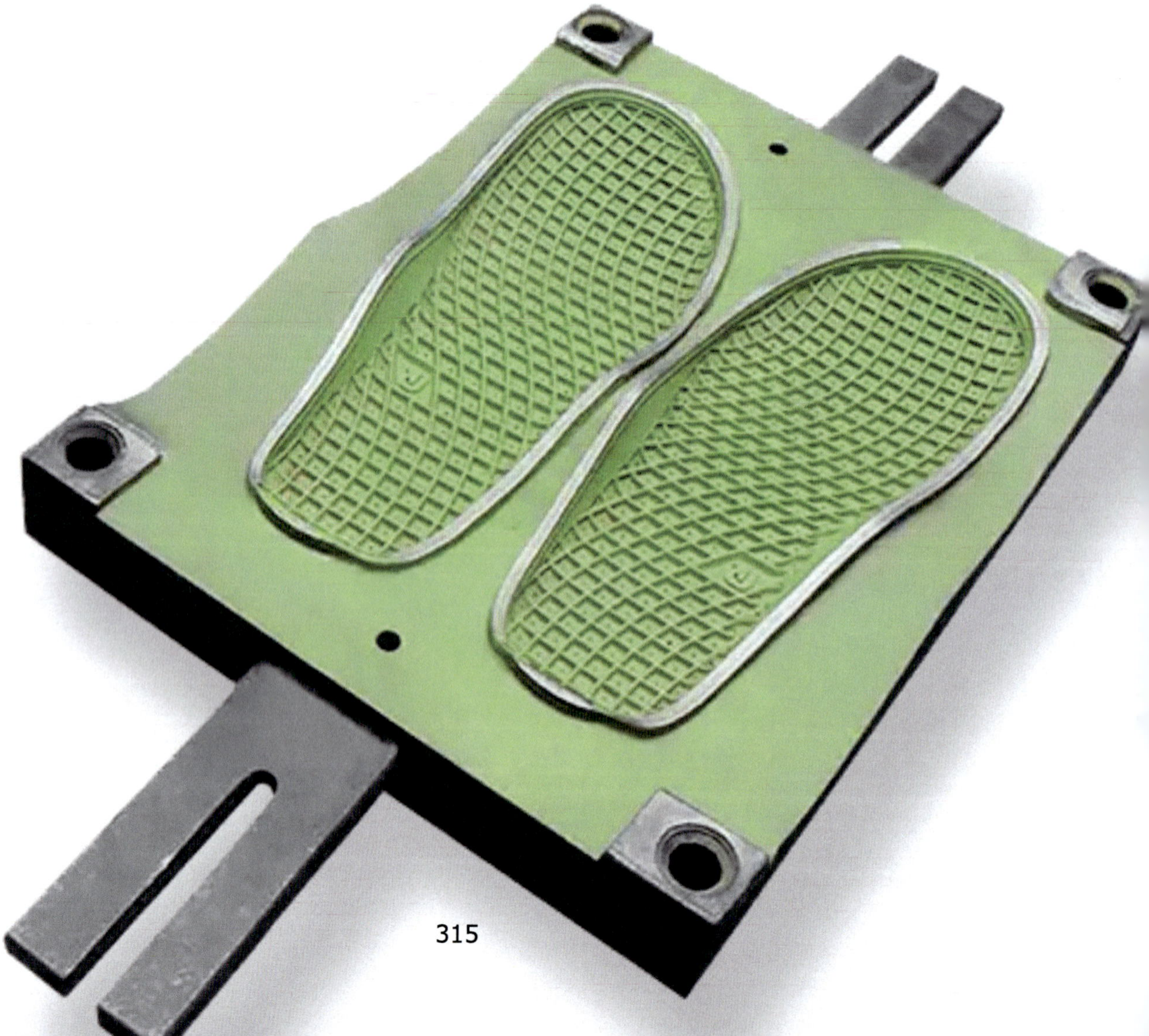

Cold pressing
EVA footbeds

Chapter 23

Silkscreen Printing Equipment

This iconic design is created with silkscreening.

Printing a design icon

Silkscreening in Footwear Production
Silkscreening, similar to t-shirt printing, is widely used for applying logos, patterns, and color details to footwear materials.

How Silkscreening Works
A polyester mesh screen with a resistant stencil is placed over the material.

Ink is forced through the screen using a squeegee, applying the design onto the fabric or synthetic surface. The printed material is then air or heat-cured to set the design permanently.

Applications in Footwear
Used in graphic sneakers like Vans or printed soccer cleats, silkscreening creates durable, colorful patterns on textiles and synthetic uppers. It can be combined with RF welding to transfer silkscreened PVC logos onto shoes.

Cost and Efficiency
Screens cost around $50 per design and can be reused. Factories may have dozens of screens for different logo sizes and patterns.

Silkscreening is Very Common
Silkscreen printing is an excellent way to apply logos to footwear. The screens themselves are inexpensive. Production capacity is almost unlimited. Color options are unlimited.

The equipment requirements for the factory are very limited. A color mixing station, large tables and a robust ventilation system is all you need.

The screen process is used to put logos directly onto the surface of shoe materials such as fabric or synthetic.

The silkscreen process is also used to create heat transfer logos. The artwork is simply printed to a heat resistant plastic film.

Printing logos

Printing heat transfer logos

Here you can see the production capacity of silkscreening. One worker can create many thousands of print logos in a single afternoon.

This logo combined with a matching radio frequency emboss tool is used to create many thousands of footbed logos.

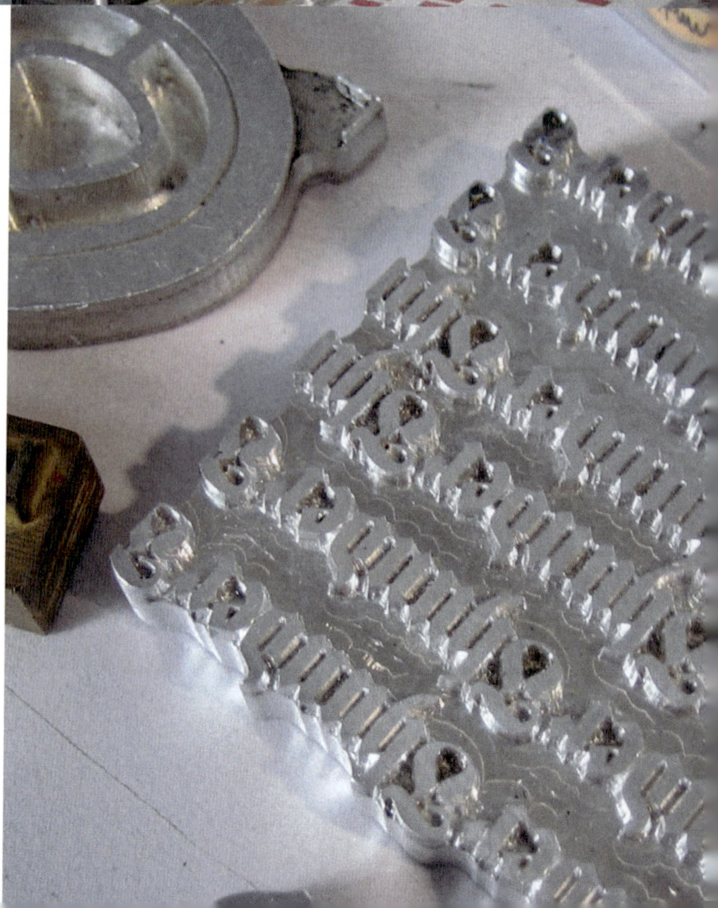

This worker is not actually printing logos. A small silkscreen fixture like this is used to create stitching guides.

In this case, the worker is using UV ink to create stitch guides that will be invisible on the finished product. When you visit the stitching department, you will see the quarter panel stitches will have the same UV light so they can follow the printed markings.

In many factories, this stitch marking function is being replaced by automatic machines.

This is a typical screen printing room in a large shoe factory. Notice the high ceiling and ventilation system.

There is an environmental downside to large screen printing operations. The air quality suffers as the ink dries, releasing solvents.

Here you can see The Shoe Dog visiting a factory making large inflatable tubes. You can see the large silkscreens are over 10 feet long and 5 feet wide. These giant silkscreens require several people working together to operate.

Yes, these are used to create one giant logo one at a time.

Chapter 24

Composite Molding For Shoes

Composite materials such as carbon fiber and fiberglass are used to create lightweight, high-strength, components in running shoes, work boots, and performance footwear.

What Are Composites?
Composites are made of fabric impregnated with resin, typically carbon fiber or fiberglass fabric.
The resin may be epoxy, polyester, or acrylic. At room temperature, these materials are flexible, but once heated, they cure into a rigid structure.

These material are often call pre-preg, meaning the reinforcing fabrics are pre-impregnated with the resin. This makes handling the material before molding much easier for the factory.

Applications in Footwear

Carbon fiber plates in running shoes
Used in high-performance models for energy return and propulsion.

Composite safety toes
Alternative to steel toe caps in industrial footwear.

Composite shanks
Provide arch support and stability in hiking and work boots.

Manufacturing Process
The composite fabric is cut into shape and placed into a mold. The mold is heated to activate the resin, solidifying the shape.

These molds are typically CNC-machined and cost around $1,000 per size.

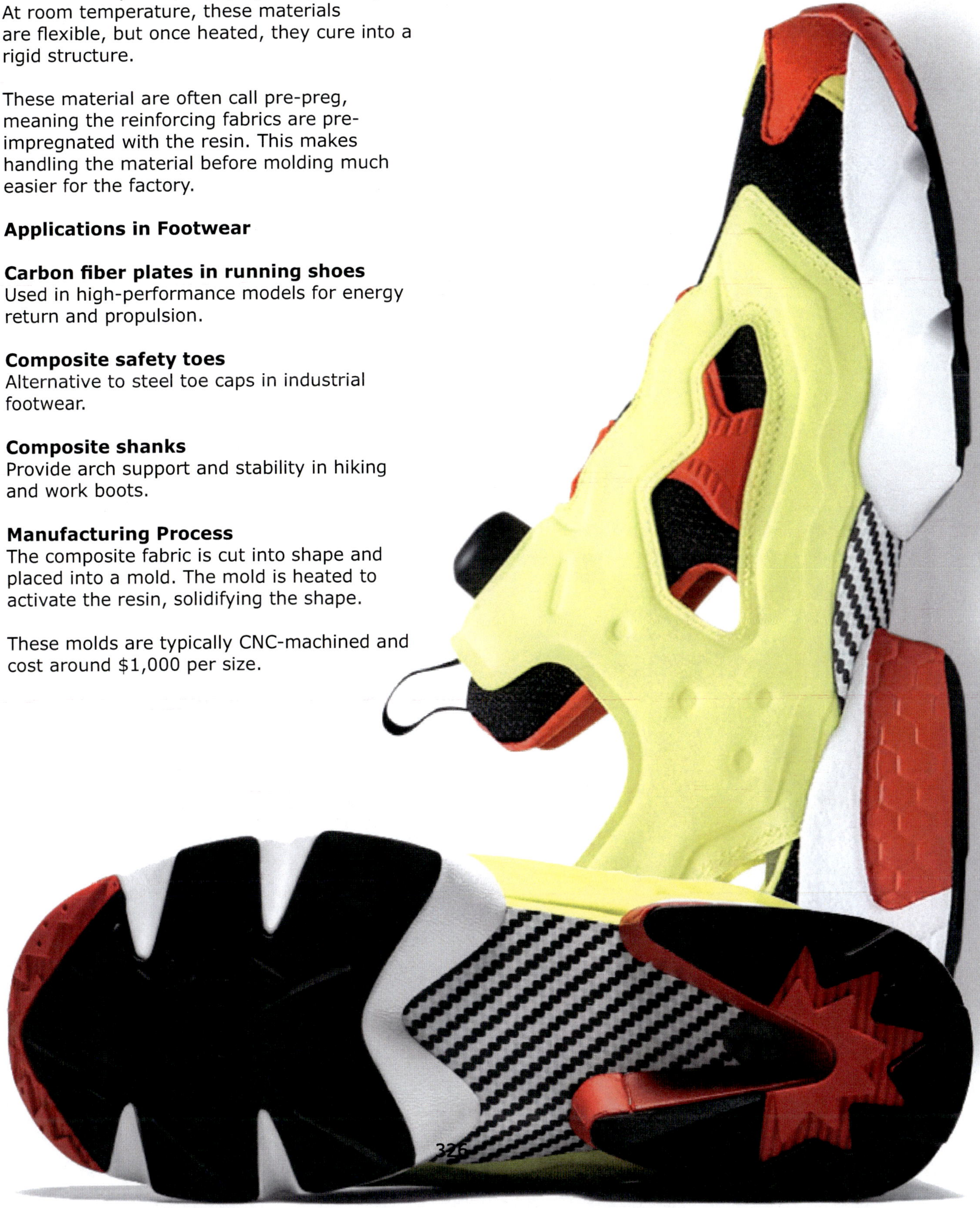

Computer cut pre-preg fabric

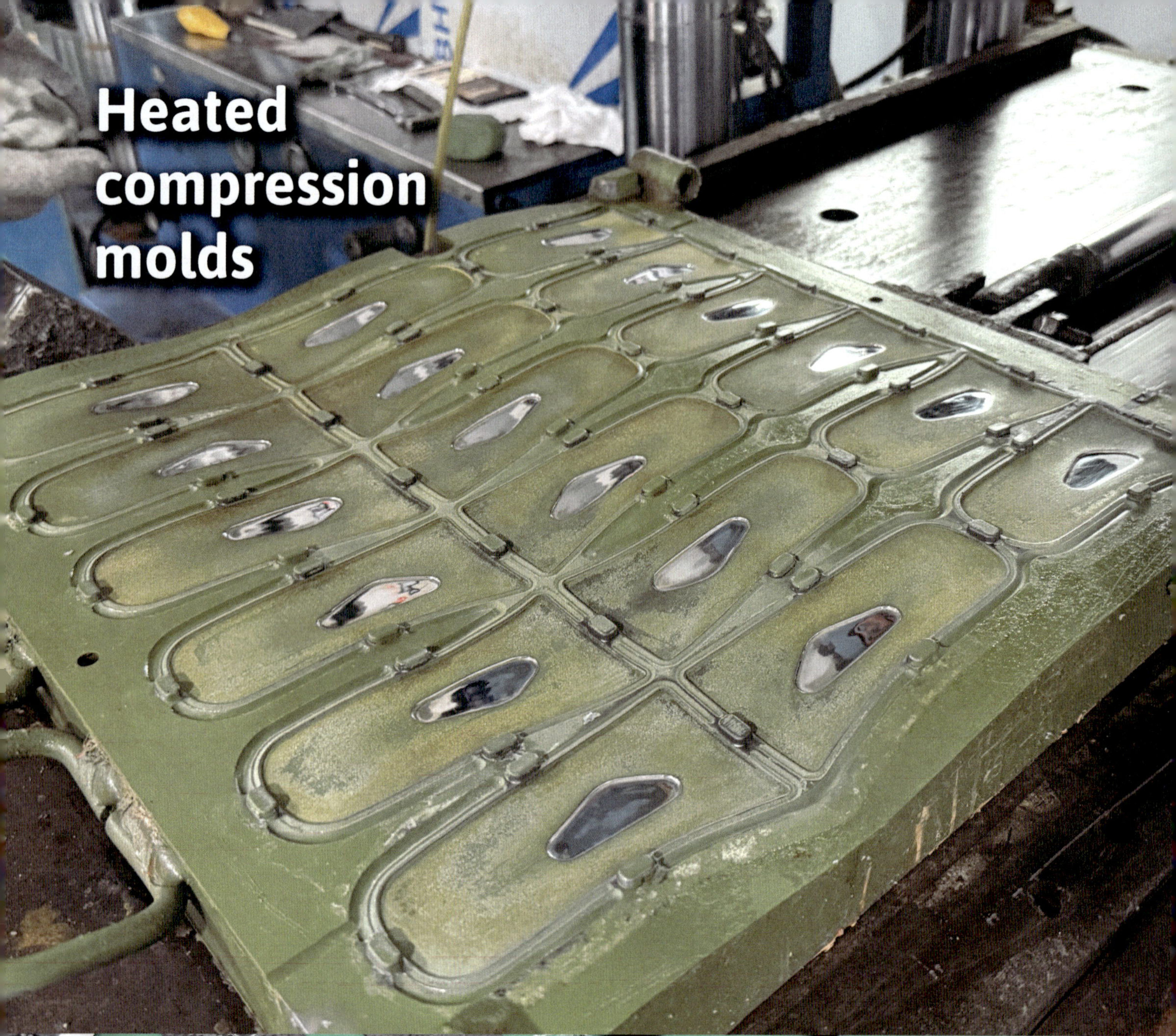
Heated
compression
molds

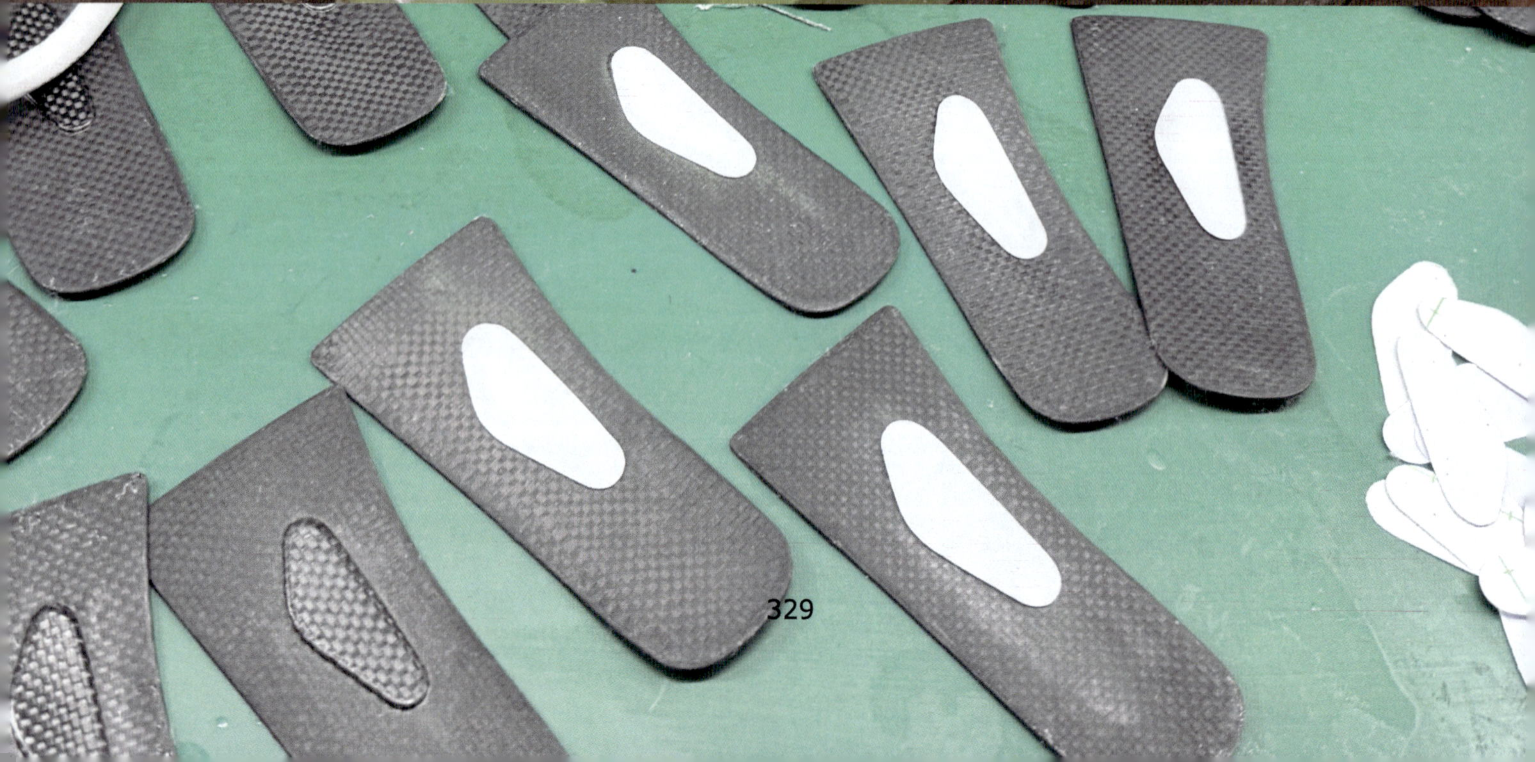

Composite safety toes made of fiberglass and carbon

Chapter 25

Vulcanized Shoe Production Tooling

This classic construction technique requires some classic equipment.

Vulcanized Footwear Manufacturing Process

Vulcanized shoes have a long-standing presence in footwear manufacturing, long known for their durability, affordability, and classic designs.

However, times have changed. With the innovations in modern production and cement process, the vulcanized is no longer the affordable staple. This classic shoe is now an indispensable fashion item.

This construction method involves assembling the shoe piece by piece, using rubber components that undergo a high-heat vulcanization process to bond them securely together and to the upper.

Key Components of Vulcanized Shoes

Rubber compression molds for the outsole
Vulcanized shoes require a compression mold to shape the bottom sole. The outsole unit is pre-molded before attaching it to the upper. This step ensures a consistent shape and texture, forming the base of the shoe.

Rubber Foxing Tape
The foxing tape wraps around the shoe's midsole and upper, securing the structure. Factories use extrusion machines to produce different textures, colors, and patterns on the tape. Custom designs require a specific roller mold to emboss logos or patterns onto the rubber.

Extrusion and Patterning
The rubber tape is either:
Extruded as a continuous strip, incorporating multiple color layers, or the rubber is rolled into sheets and cut into strips, using a textured roller mold for embossing.

Factories may add multiple stripes to the rubber before slicing it into individual tapes.

The textured waffle tape and customized patterns require precision tooling to achieve consistency.

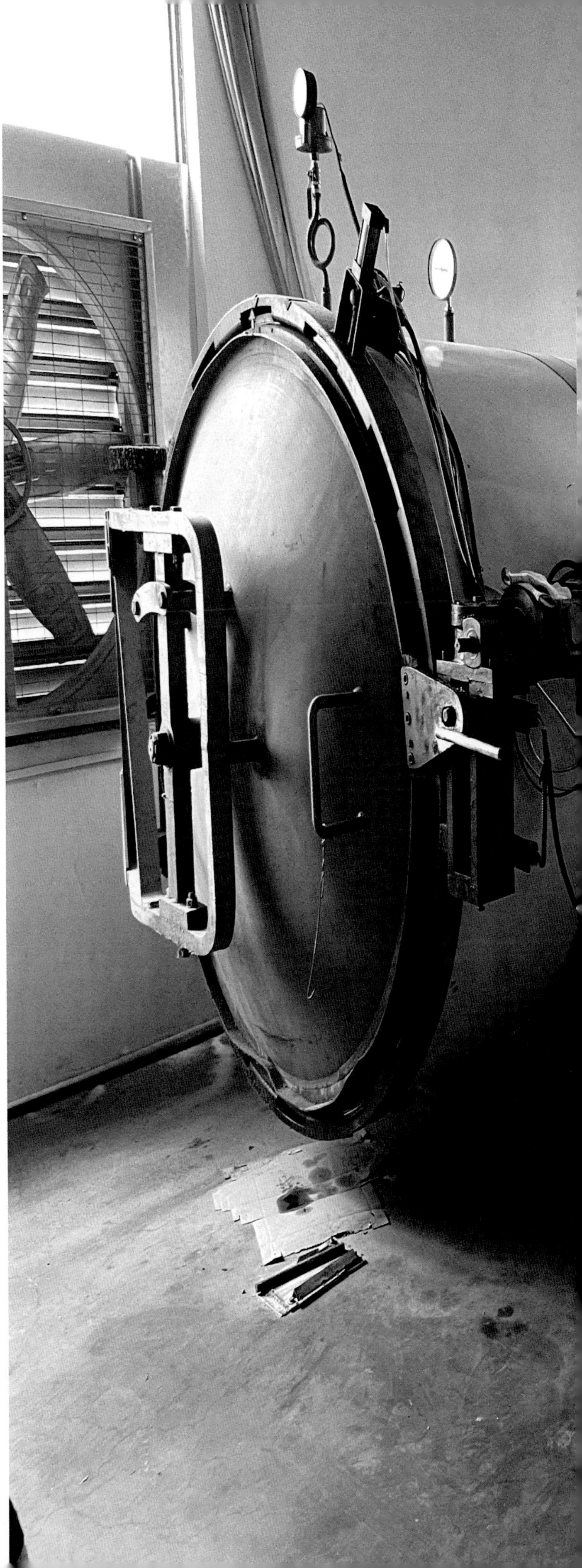

Assembly Process

Cutting and Shaping the Tape

The foxing tape is cut to the correct length and shape using scissors or a cutting die. Factories may use metal cutting dies for precise shapes, such as toe caps and branding elements.

Vulcanized Branding and Logos

Small rubber logo patches, often seen on the heel of brands like Vans or Converse, are made using compression molds. The mold presses and cures the rubber, ensuring a durable embossed logo. Factories typically create multiple logos per sheet to maximize production efficiency.

Final Assembly and Vulcanization

The upper is positioned onto the rubber outsole. The foxing tape is wrapped around the shoe, providing reinforcement.

Once assembled, the shoe is placed into a vulcanization chamber where high heat and pressure cure the rubber, bonding all components together.

Two color
foxing tape
extrusion

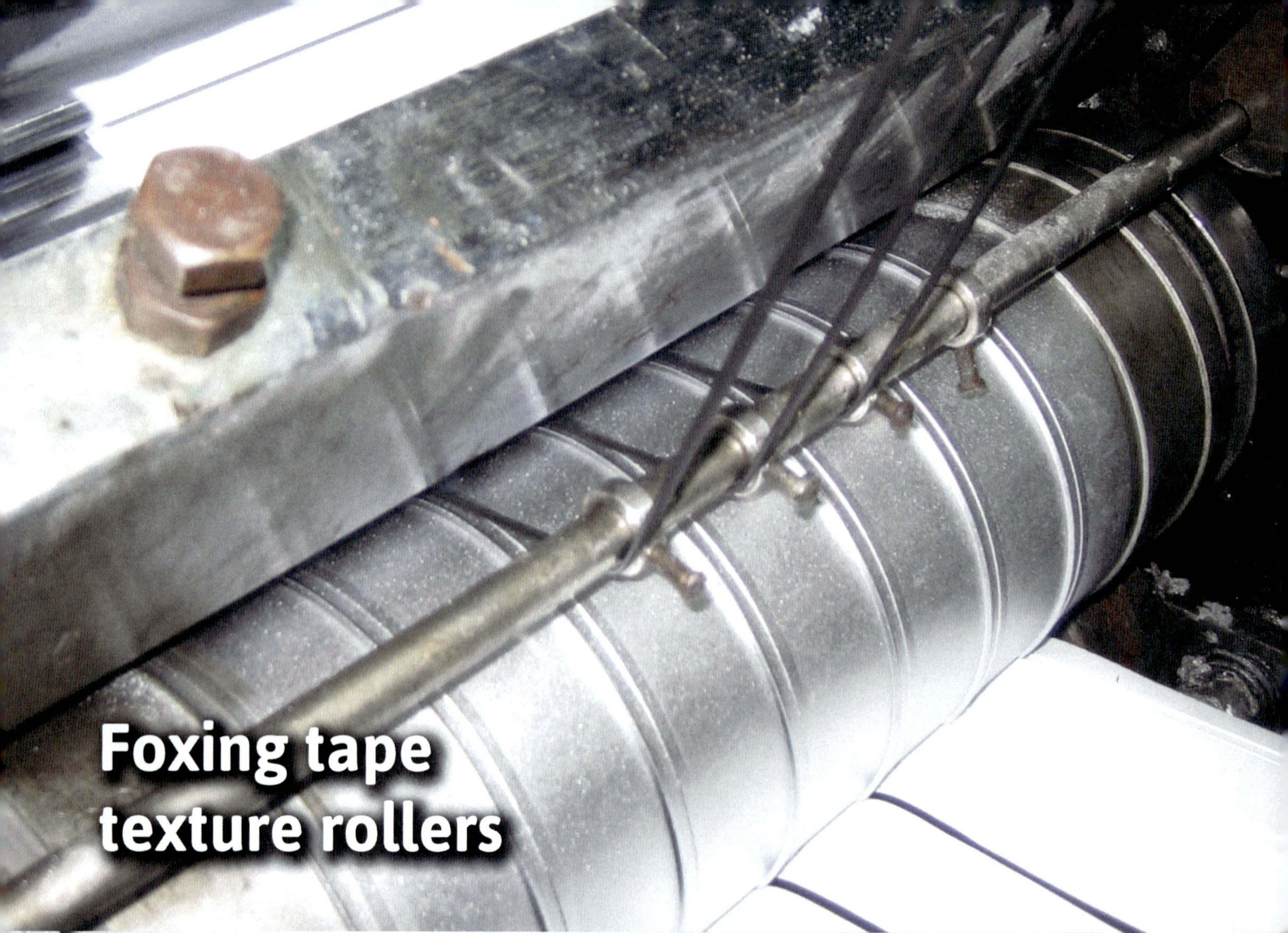

Foxing tape texture rollers

Foxing tape
5-color machine

Advantages and Challenges of Vulcanized Shoes

Advantages

Durability
The vulcanization process creates a strong bond between the sole and upper.

Cost-Effective Production
Requires relatively simple equipment compared to injection molding.

Customization
Different tape colors, patterns, and layers allow for unique design variations.

Classic Aesthetic
The timeless style appeals to a broad consumer market.

Challenges

Labor-Intensive Requires more manual labor than modern injection-molded footwear.

Raw Rubber Shelf Life Rubber components must be used quickly before they become stiff or brittle.

Not Ideal for High-Performance Footwear Less cushioning and support compared to modern EVA injection-molded shoes. Vulcanized footwear remains a staple in casual and skate shoe production, offering affordable and durable designs with classic aesthetics. While the process is somewhat traditional, it allows for high customization and cost-effective production.

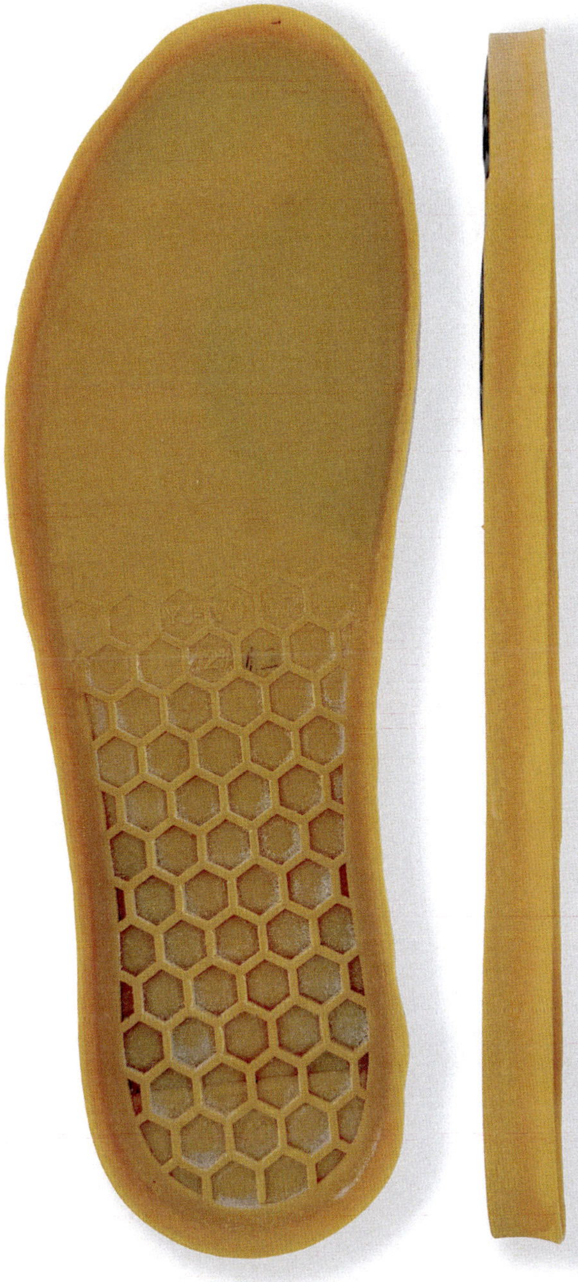

Brands like Vans, Converse, and other skate and lifestyle companies continue to rely on vulcanized construction for its iconic appeal and reliability.

Although it requires more labor and manual steps, the simplicity of the equipment and the ability to produce highly textured, custom-designed footwear, keep this method relevant in the industry.

Chapter 26

Managing Your Tooling

Tooling Management in Footwear Manufacturing

Once tooling is designed and produced, it must be properly managed to ensure its longevity and efficiency. Tooling is often a significant investment and mismanagement can lead to financial loss, production delays, and even legal disputes over ownership. We have established best practices for tooling management, supplier relationships, and maintenance procedures.

The Importance of Tooling Management

Understanding the Risks of Poor Management

Tooling is expensive and in many cases, irreplaceable. If a factory goes bankrupt, tooling might become inaccessible or lost due to unpaid debts. In some instances, tooling suppliers or mold makers may retain the equipment if they were not fully compensated. This can force brands to either pay additional costs to reclaim their tooling or start from scratch with a new design.

To avoid such scenarios, brands must establish clear agreements with subcontractors, maintain accurate records, and frequently inspect their tooling.

Where is your tooling?

Where should it be?

Identifying Tooling Location and Ownership
Since most shoe factories do not manufacture rubber, EVA, or injection-molded components in-house, tooling is often held at subcontractor facilities.

Key questions to ask include:

Where is the tooling located?

Who has control over the equipment?

What is the relationship between the brand, the shoe factory, the agent, and subcontractors?

How is the tooling stored and maintained?
It is essential for brands to maintain direct relationships with subcontractors handling their tooling. Regular visits to these facilities help ensure that tooling is properly stored and maintained.

Typical Supplier Relationships

It's typical for the brand to pay the shoe factory for a piece of equipment. In this example, the cutting dies. Even though the shoe factory does not make the cutting dies, they subcontract the production out to a specialist.

In the case of cutting dies, it's unlikely that the brand itself would have a direct relationship with a cutting die factory. As the shoe brand is unlikely to be involved in the design of the cutting dies.

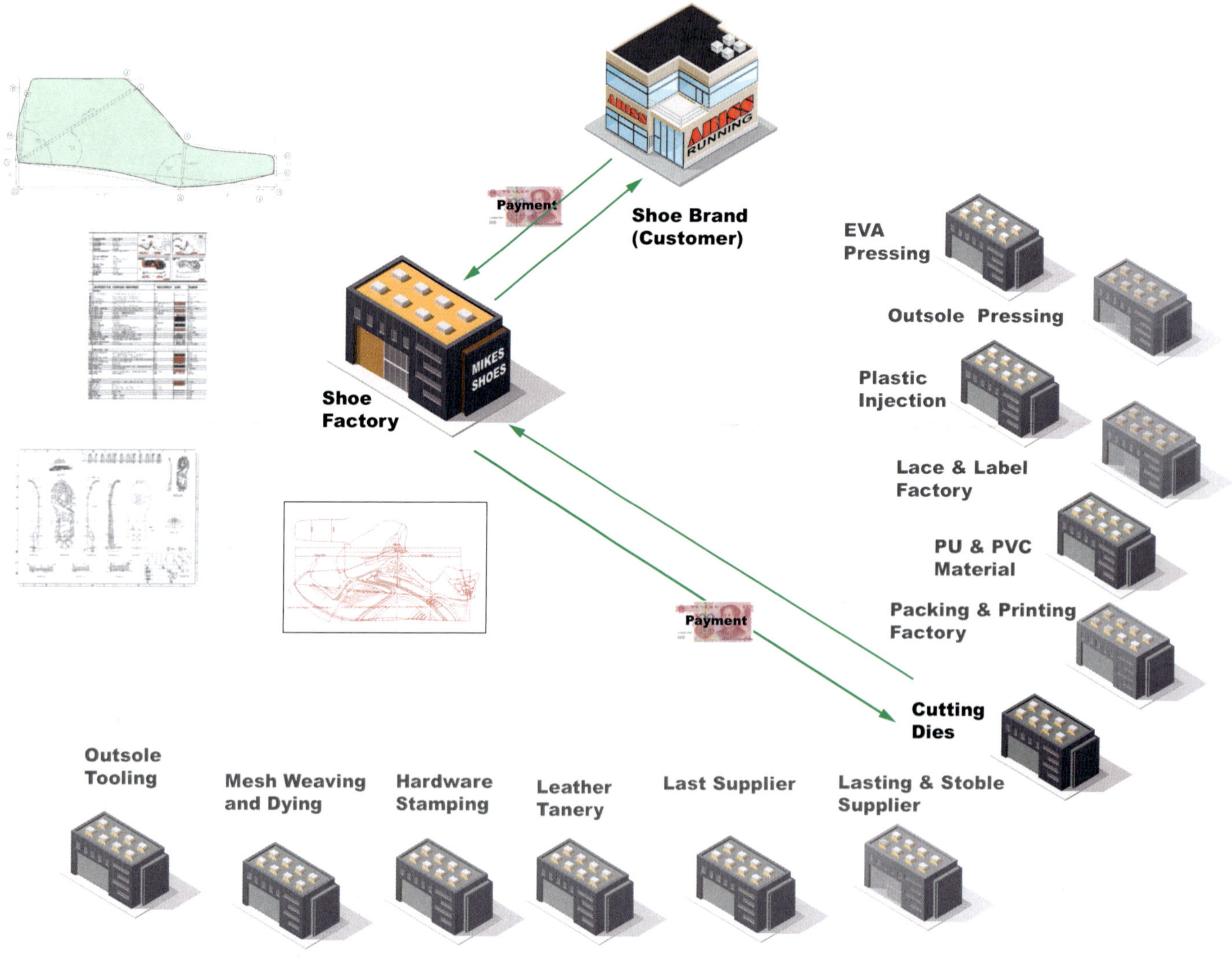

Typical Supplier Relationships

When it comes to shoe last development, it's common for a brand to have a direct relationship with the last supplier. For large brands that support production in multiple countries it's common for their last supplier to also have branch offices in multiple countries.

In this case, the last supplier is nominated to supply the last for all of our brands partner factories. The shoe factory might purchase the last on the brand's behalf.

This allows the brand to supply the same last designs to competing shoe factories. This gives the brand control of a critical asset.

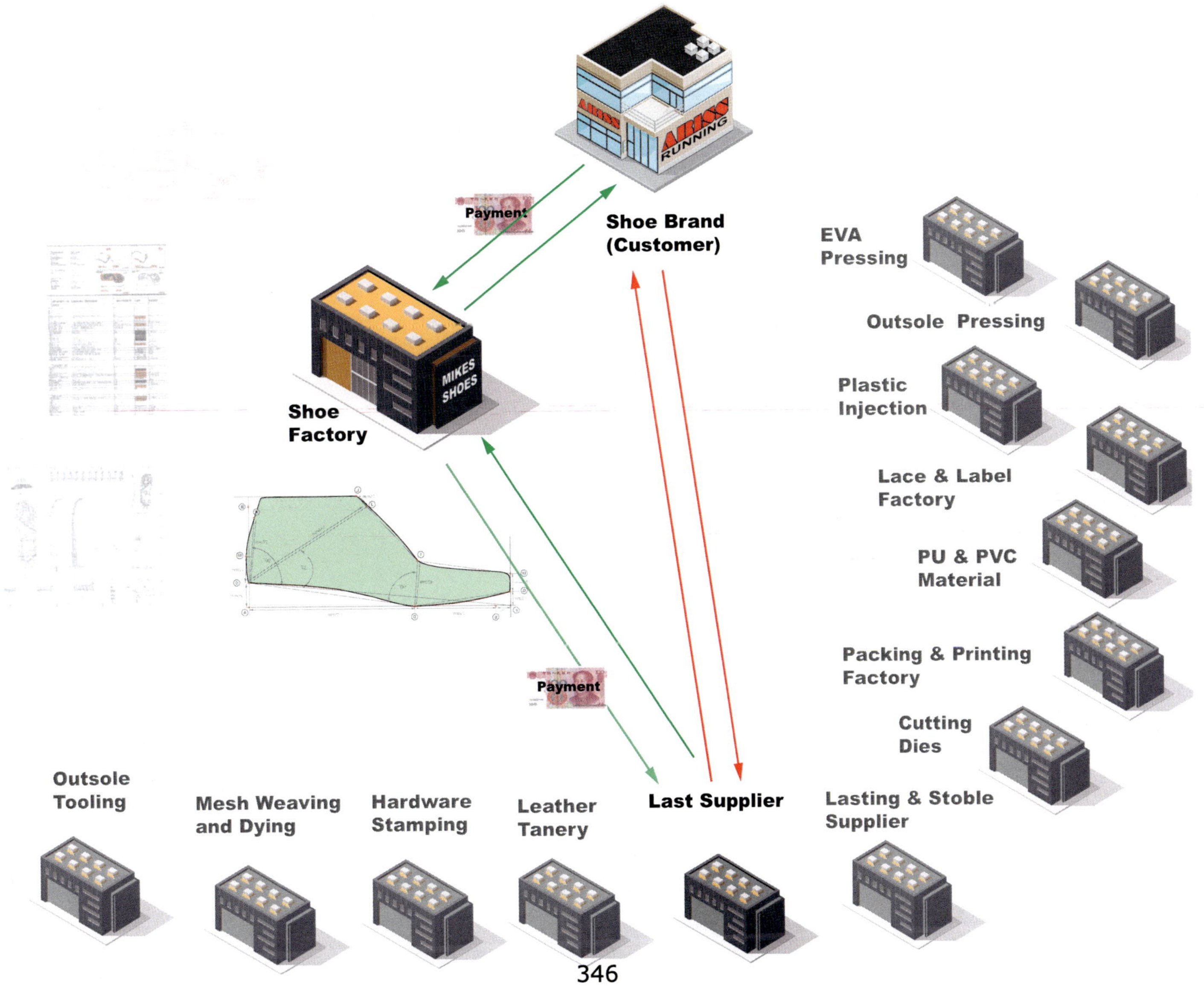

Typical Mold Supply Relationships

In this case, the shoe brand and the shoe factory work together to develop the mold designs. After the tooling factory creates the molds, the equipment is passed to the parts supply subcontractors.

The molds are not located at the shoe factory at all. As a brand, if you would like to see your parts in production, you may need to ask the shoe factory where the parts are being made.

This is not ideal as a brand. You should know where all of your parts are coming from.

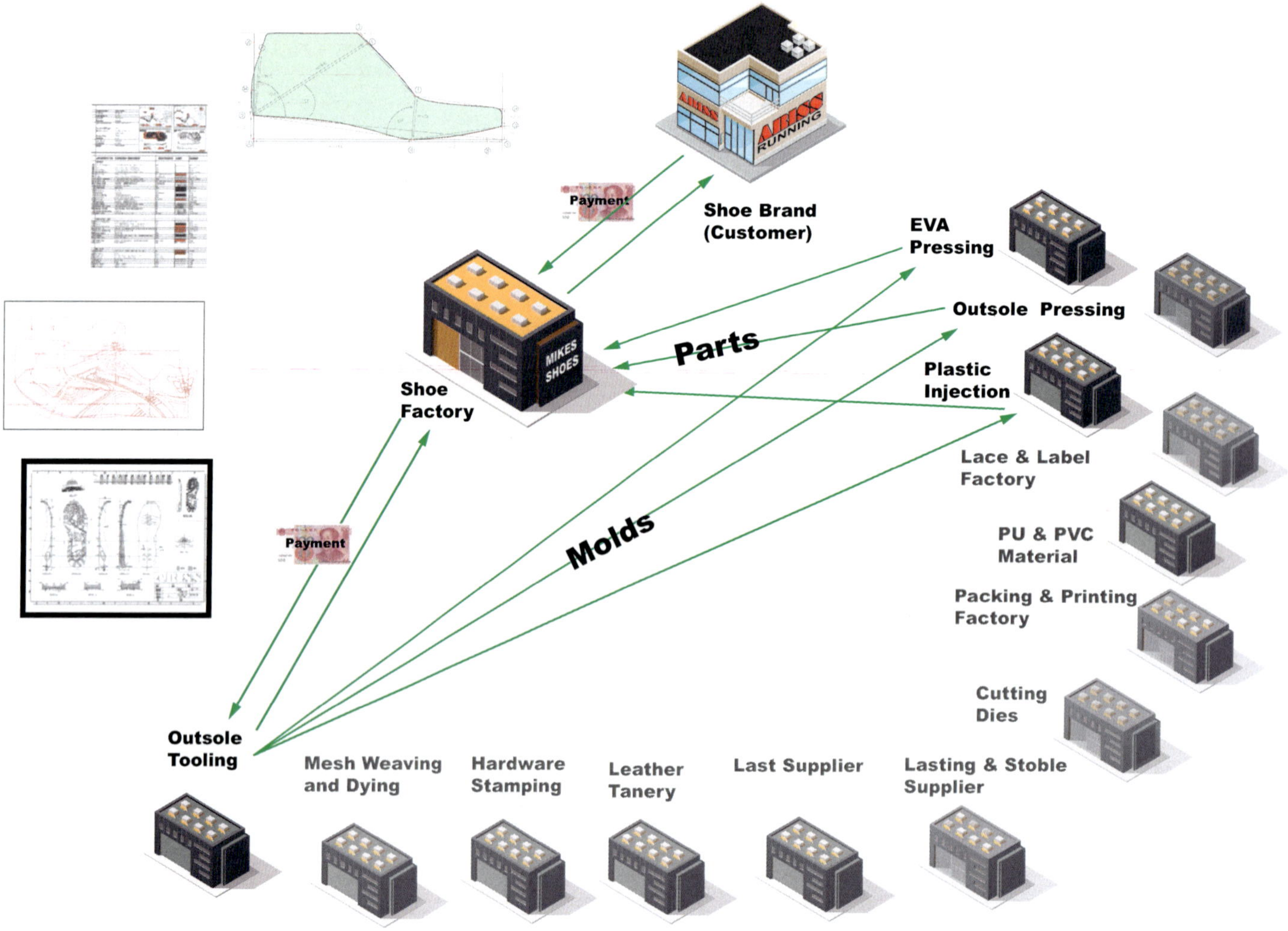

Tooling Factory Relationships

Just as the shoe factory itself does not produce all of the sub components for the shoes, neither does the outsole tooling factory. The outsole tooling may require several subcontractors to finish your molds. Most notably, metal casting, mold texture, and Teflon coating may be done off-site from the main shoe factory.

Once the tooling is all complete, the outsole tooling factory will send it to the shoe factory.

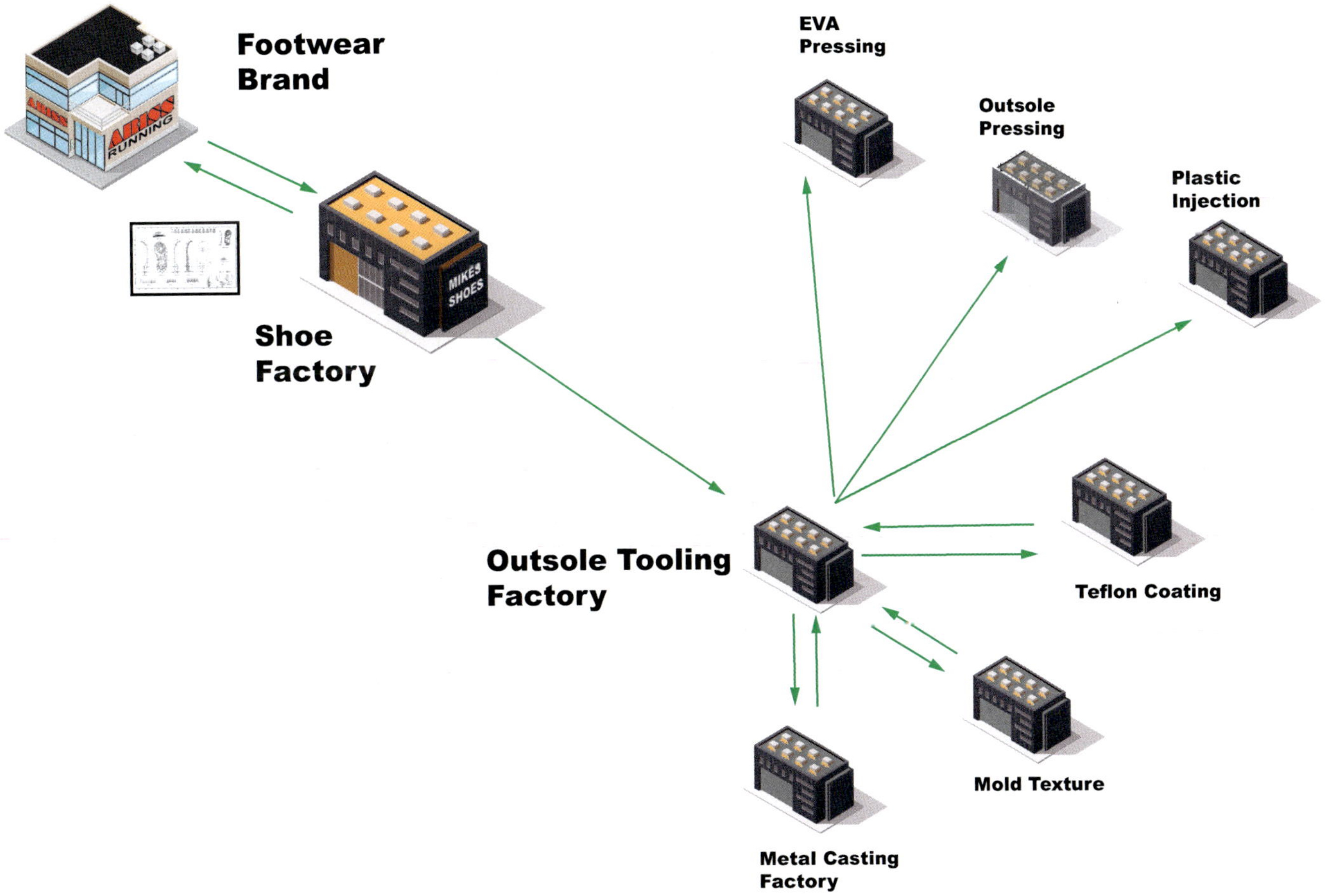

Better Supplier Relationships

The best way to manage your equipment at subcontractors is for you to have direct relationships with all of the subcontractors. Of course, you have a relationship with your shoe factory, but you should also know the outsole factory, and you should take time to visit all of your part producing subcontractors.

First, so you can inspect your molds and the parts. And second, so you can maintain relationships with all of the subcontractors independent of the tooling factory and the shoe factory.

Remember, in the future, you may want to send parts to a different shoe factory.

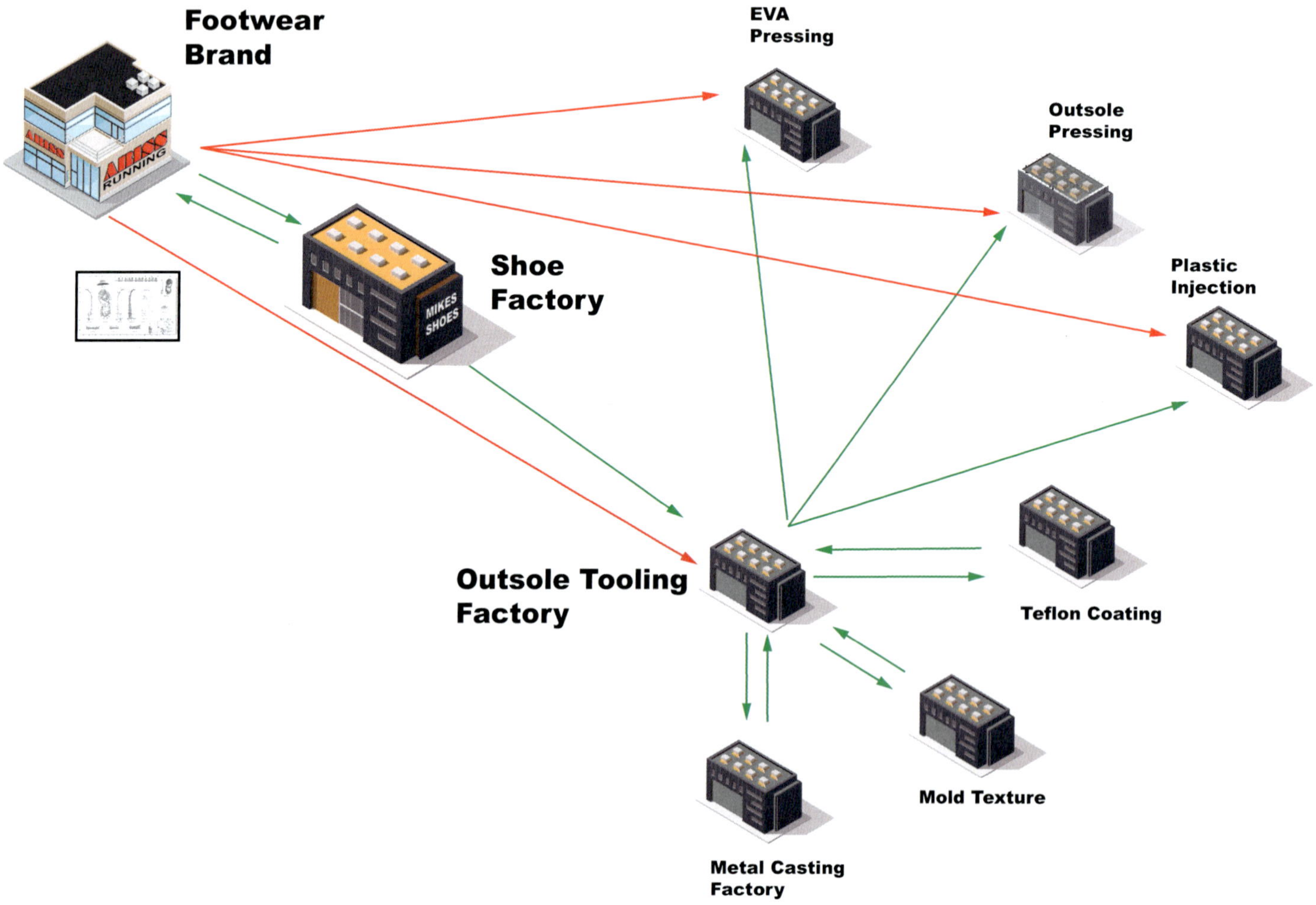

Sourcing Agents May Not Be Working in Your Best Interests

Some brands operate through agents who act as intermediaries between the brand and manufacturers. While agents provide convenience, they can also limit a brand's control over its tooling and supplier relationships.

If an issue arises between an agent and a supplier, brands may be left without access to their equipment.

To mitigate this risk, brands should insist on visiting subcontractors directly. If an agent refuses to facilitate such visits, it may indicate a lack of transparency, prompting the brand to reconsider its partnerships.

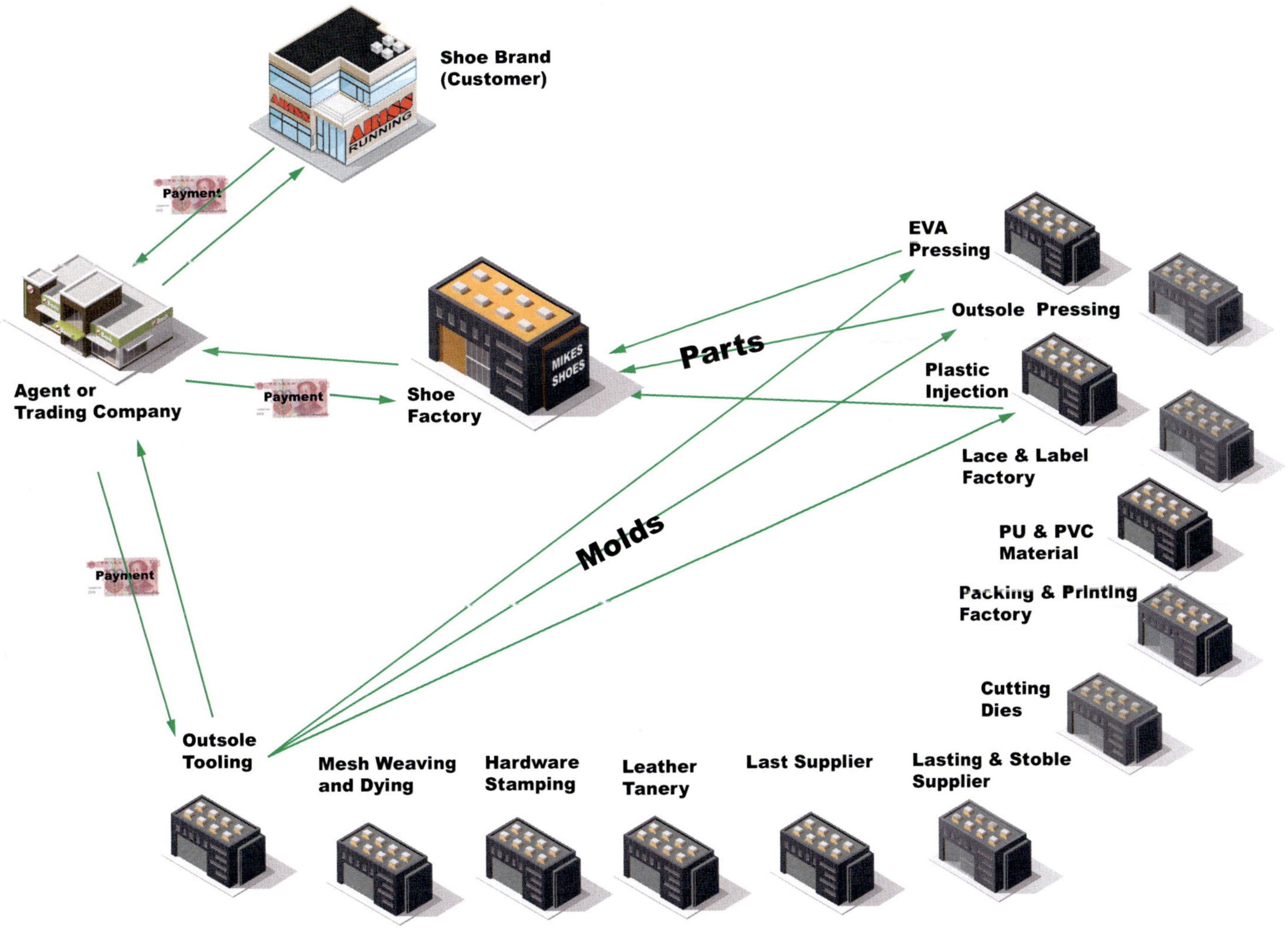

Chapter 27

Controlling Your Tooling Inventory

All of your tooling should have ID plates and you should maintain a complete inventory. You need to know where your equipment is and make sure to check it once a year.

If tooling is obsolete, pay to scrap it. If you are not sure, move it to a safe place that you can control.

Tooling Inventory and Identification

Labeling and Tracking Tooling

Even small brands accumulate significant amounts of tooling, making inventory management crucial.

Every piece of equipment should be clearly labeled with the brand's name, a unique identification number, or a visible tag on the exterior of the mold.

This labeling ensures that tooling can be located quickly and prevents unauthorized use by other factories. Many large companies conduct annual tooling audits, physically inspecting each mold to verify its condition and location.

Proper Tooling Storage

Proper storage prevents deterioration and ensures tooling remains in usable condition. However, not all factories prioritize storage quality. Poorly managed molds may be left outside and exposed to rain and rust or piled haphazardly, making identification difficult. Tooling may also be stored in unsanitary conditions, leading to defects in production.

To prevent this, brands should visit storage areas and enforce proper maintenance protocols.

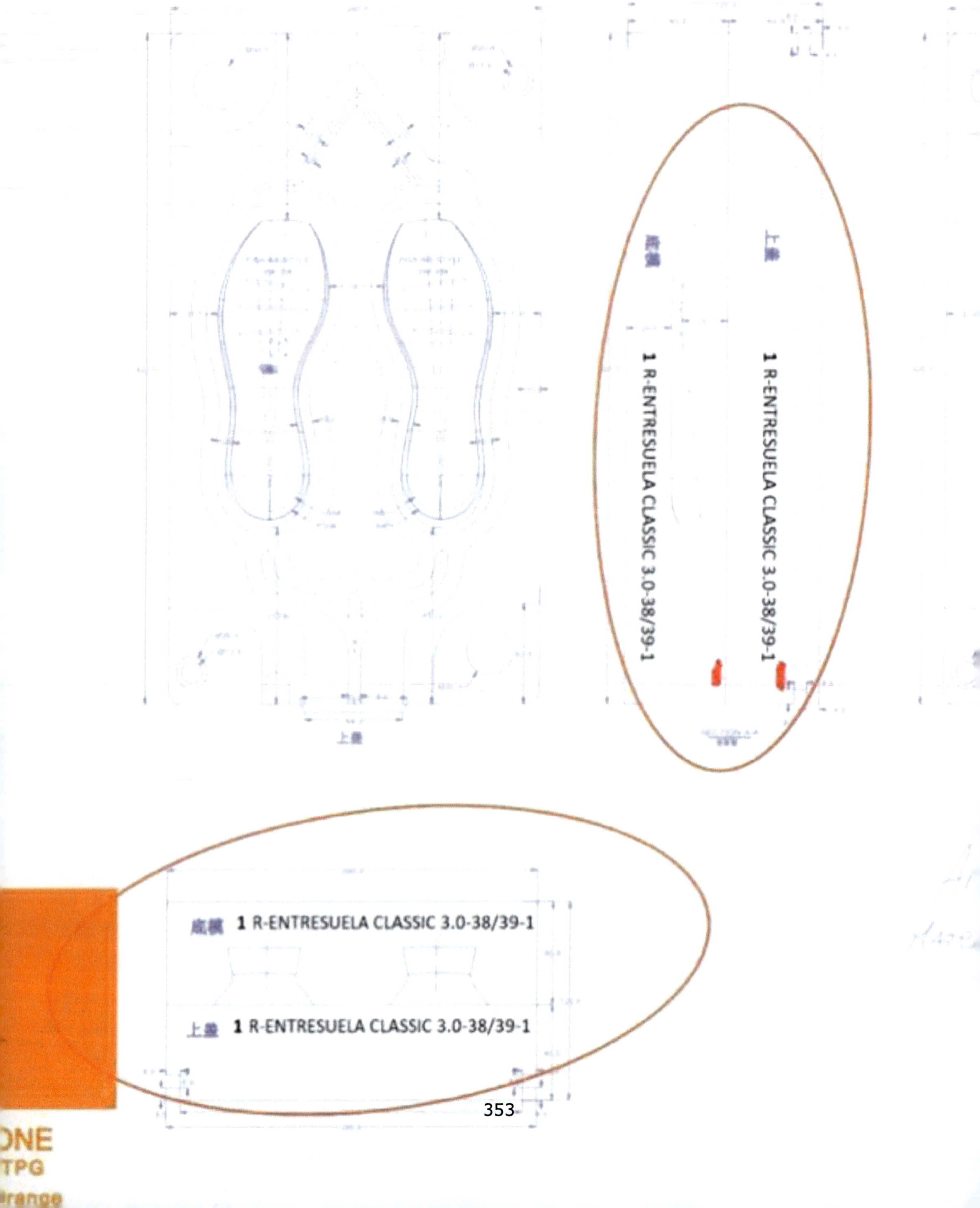
1 R-ENTRESUELA CLASSIC 3.0-38/39-1
1 R-ENTRESUELA CLASSIC 3.0-38/39-1
1 R-ENTRESUELA CLASSIC 3.0-38/39-1
1 R-ENTRESUELA CLASSIC 3.0-38/39-1

Cutting ID numbers with a vintage pantograph

Laser cutting
ID numbers

MS75A
S2772
45A
S2772

Where are your molds?

Chapter 28

Footwear Tooling Care & Repair

Tooling Maintenance and Repairs

Regular Maintenance to Prolong Tool Life

High-use molds require consistent maintenance to prevent defects in production.

Leading factories proactively maintain their molds by:
Scheduling regular inspections based on production cycles.

Cleaning and resurfacing molds to prevent defects.

Repairing worn parting lines and damaged surfaces.

Some factories invest in dedicated maintenance departments to keep tooling in optimal condition. Without proactive care, worn molds can lead to product defects, increasing production waste and costs.

Common Tooling Issues and Repairs

Tooling naturally wears down over time.

Typical maintenance procedures include:
Welding and machining: adding material to restore worn mold sections.

Replacing inserts: swapping out damaged sections instead of rebuilding entire molds.
.
Re-coating with Teflon or other finishes: preventing adhesion and ensuring smooth part release.

Factories that do not prioritize maintenance may rely on makeshift fixes such as using aluminum tape to seal gaps in worn molds. While this may work temporarily, it is not a long-term solution.

Avoid Common Pitfalls in Tooling Management

Preventing Tooling Neglect

Molds that are left unused for extended periods often become obsolete or deteriorate due to poor storage conditions. Brands should regularly review their tooling inventory and decommission old molds properly.

Ensuring Proper Disposal of Obsolete Tooling

If a mold is no longer needed, brands must oversee its disposal to prevent unauthorized use. If left unattended, outdated tooling could be repurposed by unauthorized manufacturers, potentially leading to counterfeit products.

Auditing Cutting Dies and Other Consumable Tooling

Unlike molds, cutting dies cannot be resharpened—once dull, they must be replaced. Rusty or worn out dies result in imprecise cuts leading to quality issues. Brands should monitor the condition of cutting dies and replace them as needed.

Effective tooling management

Effective tooling management is critical to ensuring production efficiency, cost control, and product quality.

Brands must take proactive steps to:

Track and inventory all tooling to prevent loss or misuse.

Develop relationships with subcontractors to maintain control over their assets.

Implement regular maintenance programs to extend the lifespan of molds and cutting dies.

Ensure proper storage and disposal of obsolete tooling to maintain security and avoid counterfeits.

By following these best practices, brands can protect their investments, minimize production risks, and maintain high-quality standards in footwear manufacturing.

The Teflon is wearing off the parting line. This will require some work but it’s not effecting the main mold cavity.

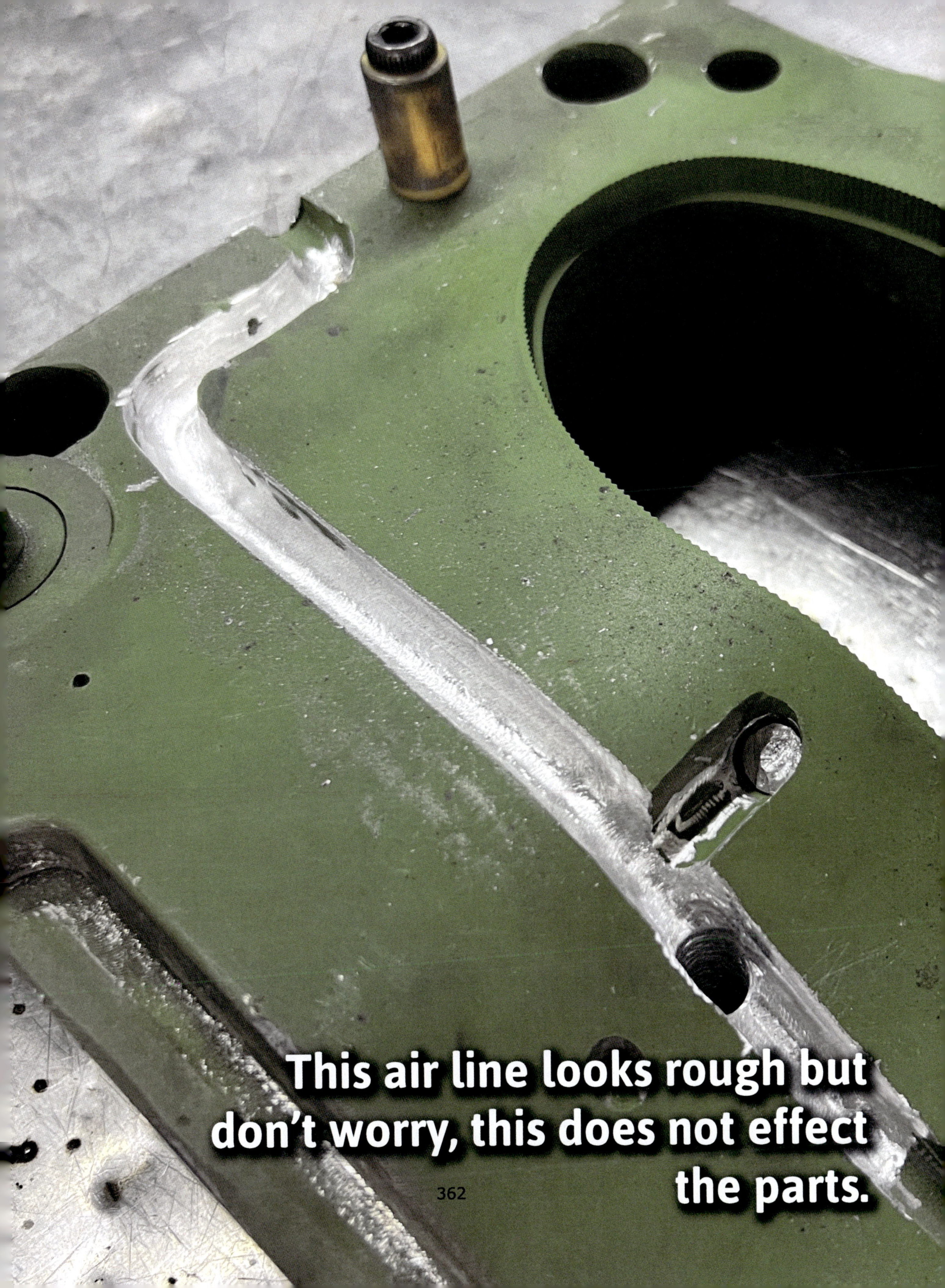

This air line looks rough but don't worry, this does not effect the parts.

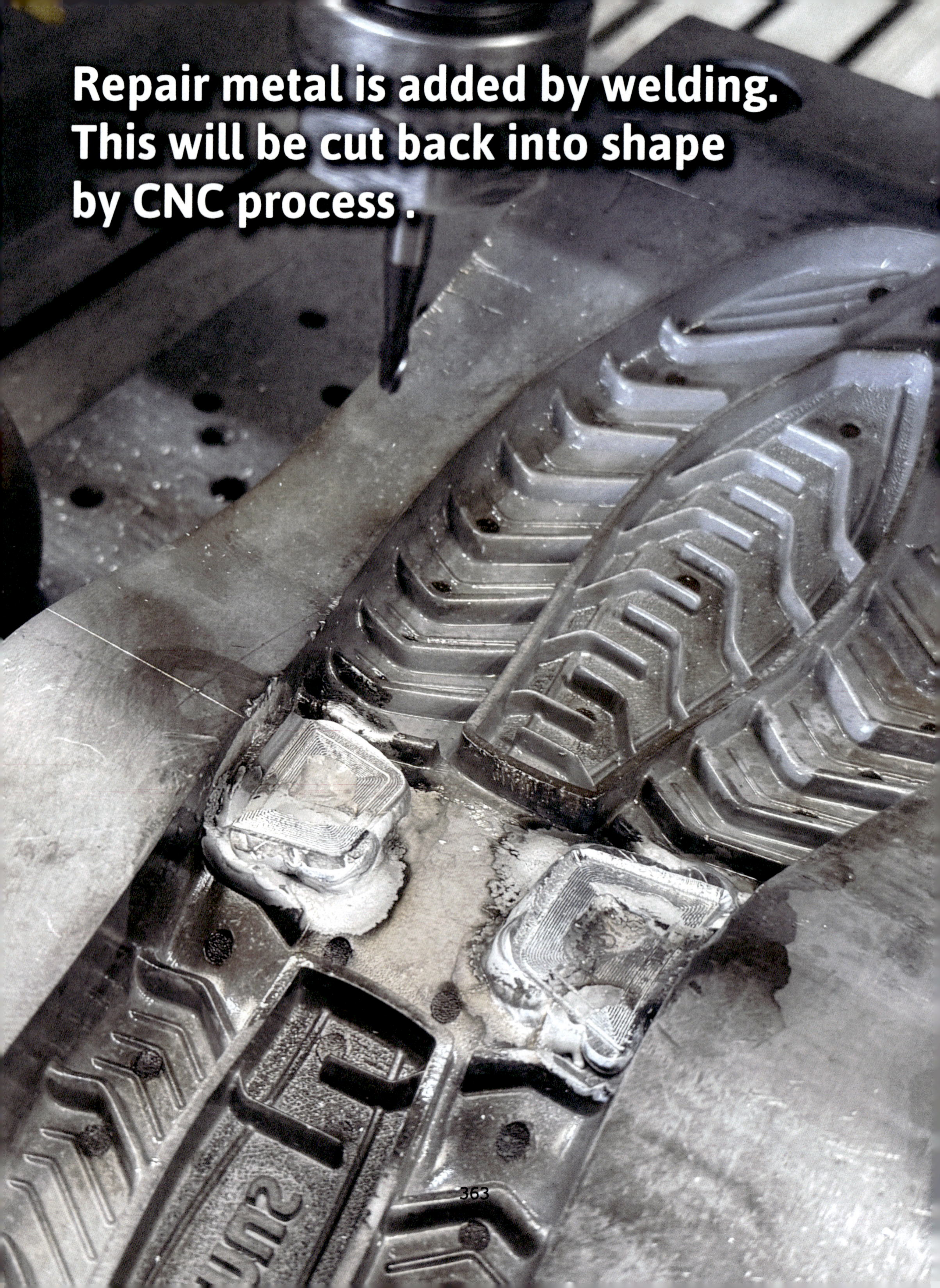

Repair metal is added by welding. This will be cut back into shape by CNC process .

This injection EVA tool desperately needs a good cleaning before we can assess the damage.

We better look inside!

Yes! This really was one
of my tools.

Poor storage or
obsolete tooling graveyard?
In either case you
don't want your tooling here.

Workers are checking the mold parting plane.

This a good thing to see.
An entire size run of
equipment
getting regular maintenance.

Repairing a worn out
parting line.

Repair metal is added by welding. This will be CNC cut back into shape.

These can't be saved.
It's recycling time.

We hope this is a
temporary repair.

Yes, shoe lasts do wear out.

These are ready to be remelted and recycled .

Chapter 29

Footwear Without Tooling

3D Printing in Footwear Manufacturing
As we explore footwear tooling, it is important to consider advancements in manufacturing that reduce or eliminate traditional tooling. While many shoes require extensive tooling, emerging technologies, such as 3D printing, offer alternatives that reduce reliance on conventional molds and presses.

adidas

3D Printing and its Role in Footwear Production

Adidas Alpha Edge and Carbon 3D Technology

The Adidas Alpha Edge is an example of footwear incorporating 3D printed midsoles using Carbon 3D printing technology. While this process eliminates traditional tooling for midsoles, the shoe still requires upper conditioning tools, welding tools, heel counters, and toe formers. Thus, while 3D printing reduces some tooling requirements, it does not entirely eliminate them.

Carbon 3D Printing Process

Carbon 3D, a pioneer in flexible TPU 3D printing, developed a laser-based process that creates soft, pliable, and durable components within a liquid polymer bath. This innovative method allows for intricate midsole structures without the need for traditional molds.

The polymer based 3D printing machine, primarily used in high-tech manufacturing hubs such as Ho Chi Minh City, Vietnam, enables the production of midsoles with superior cushioning and flexibility. However, the material cost remains high, making this technology more suitable for high-end or limited-edition footwear rather than mass production.

Applications of 3D Printing in Footwear Design and Prototyping

Rapid Prototyping for Injection-Molded Shoes

3D printing is particularly useful for prototype development before committing to full-scale production. For example: if a brand is developing an injection-molded EVA shoe, a wearable prototype can be created using 3D printing at a fraction of the cost of traditional mold fabrication.

Instead of spending $3,000 or more on an initial mold, a 3D-printed prototype can be produced for a few hundred dollars, allowing designers to test functionality before mass production.

Fully 3D-Printed Footwear

Some brands are experimenting with producing entire shoes using 3D printing, including sandals and slides with fully integrated midsoles and uppers and custom footwear components, such as high-performance insoles or modular midsole units.

While the feasibility of fully 3D-printed shoes for mainstream markets is still evolving, the technology offers promising applications for custom footwear, medical orthotics, and high-performance sports shoes.

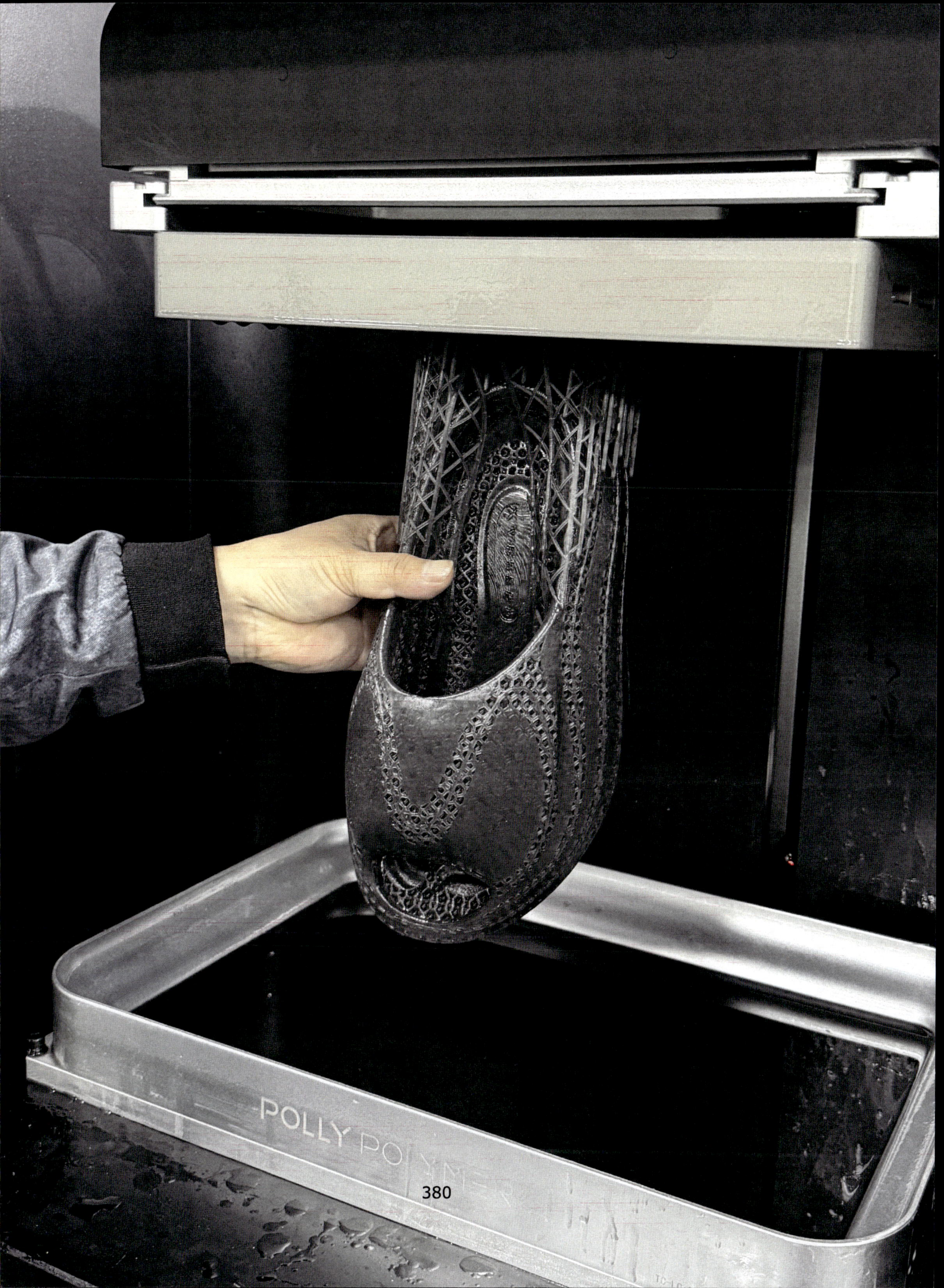
POLLY

Advancements in 3D Printing for Footwear

Expansion of Color Options

Early 3D printing technologies were limited in color variety. However, modern advancements now allow for a broader range of colors and material finishes, enabling greater customization and branding possibilities.

Integration of Upper and Outsole Printing

Recent developments in multi-material 3D printing allow for simultaneous printing of both the upper and outsole, reducing assembly steps and increasing production efficiency.

Conclusion

3D printing is revolutionizing footwear manufacturing by reducing tooling costs, accelerating prototyping, and enabling new design possibilities. While traditional tooling remains necessary for many shoe components, 3D printing offers an alternative approach that is particularly beneficial for customized, high-performance, and experimental footwear designs.

As the technology continues to evolve, its role in mainstream footwear production will likely expand, offering brands greater flexibility, efficiency, and creative freedom in shoe manufacturing.

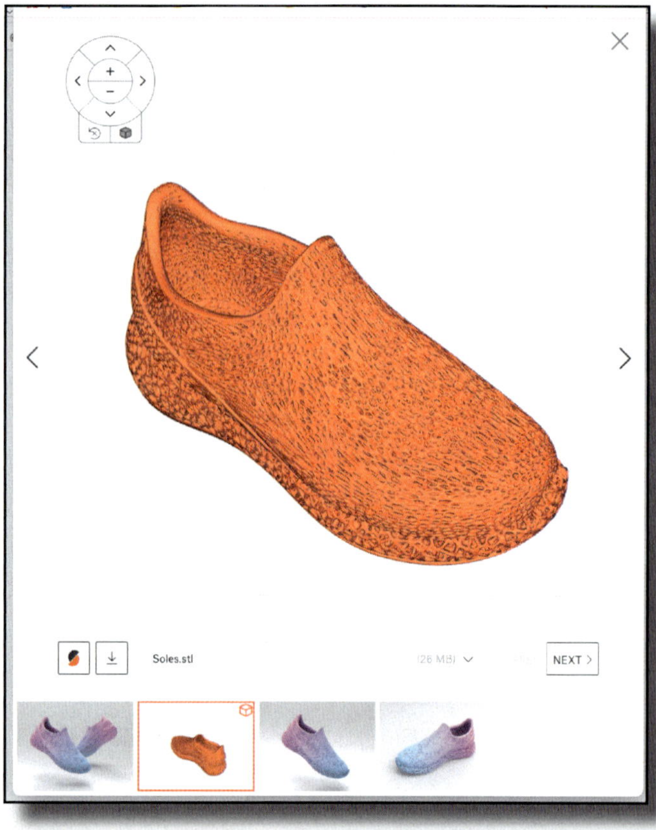

Chapter 30

Footwear Tooling Requirements & Costs

Connecting the Dots

Now that we have an understanding of the tooling available to us to make a modern sports shoe, we can review a drawing and build a list of required tooling. Once we have the design requirements, we can consider the amount of tooling we need to mass produce the entire size run.

While I have never worked for Nike, Adidas, or Li-Ning, we can use their designs as models for building up our list tooling requirements.

Alpha Edge 4K

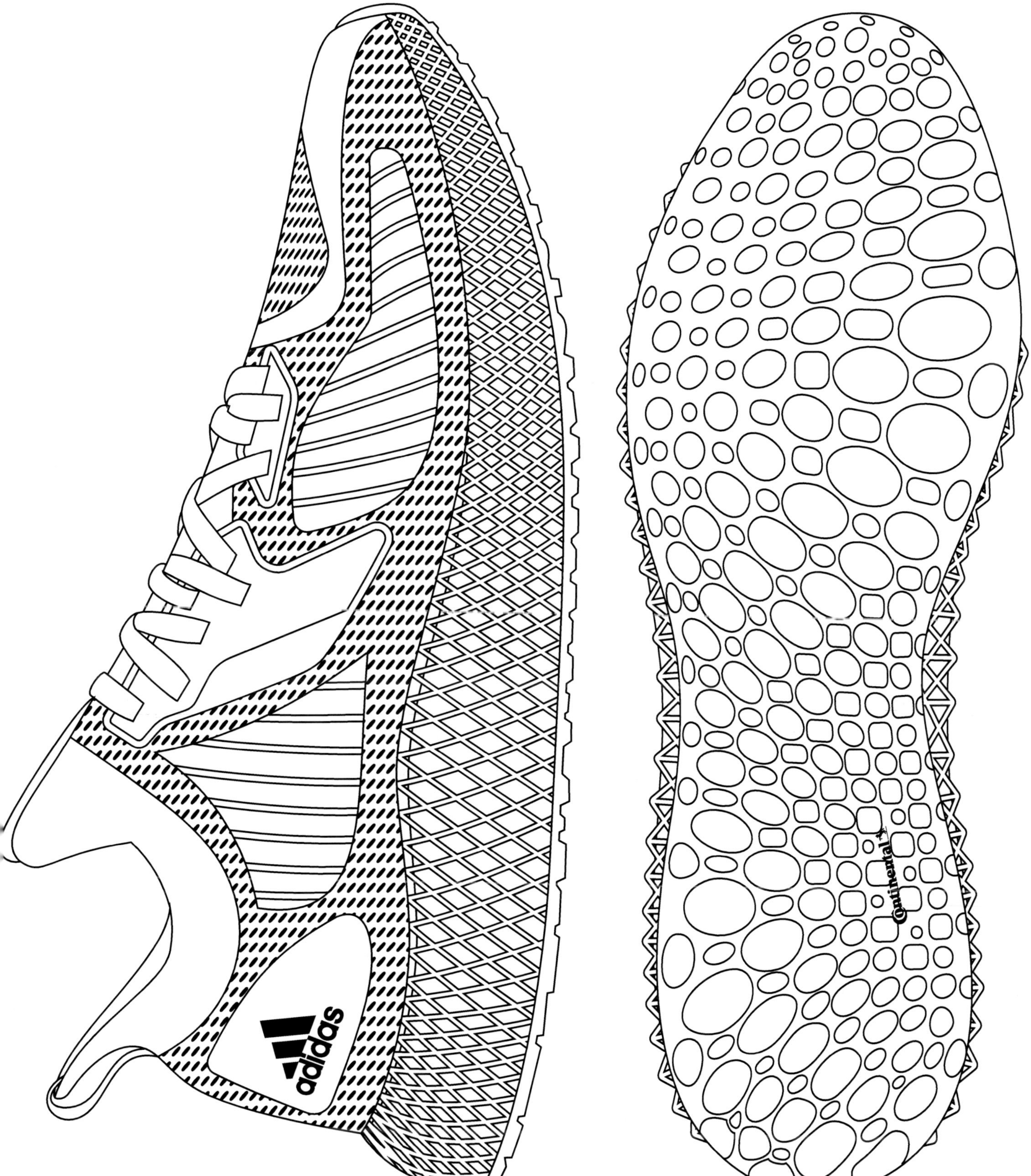

Alpha Edge 4K

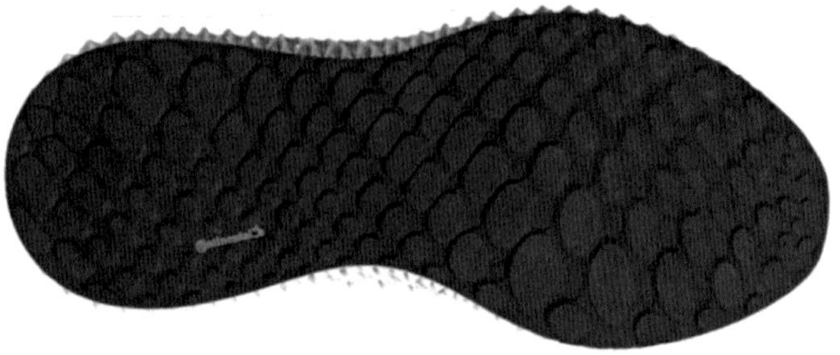

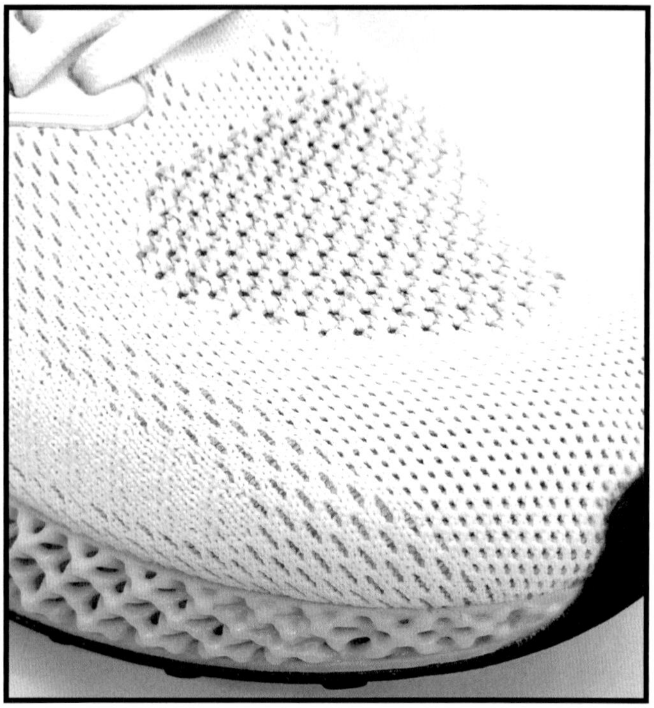

Style: Running / Technological Experimentation
Brand: Adidas
Model: Alpha Edge 4K
Introduction: 2020
Construction: Strobel lasted
Assembly: Cold cement
Sole unit: 3D printed matrix / rubber

The Adidas Alpha Edge 4K is a technical experiment combined with conventional construction techniques.

4K knitted upper
Requires programming and shaping fixture

TPU reinforcements
RF welding tooling + logo screen print
Cutting die

Outsole
3D printed matrix midsole
Rubber pressing mold
Footbed cold press mold
PU pressing pad

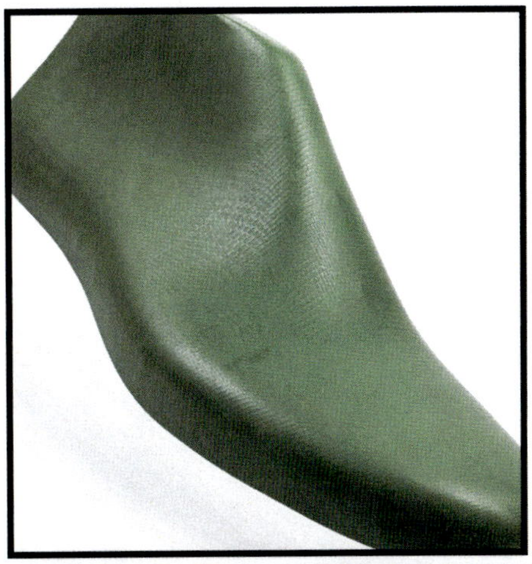

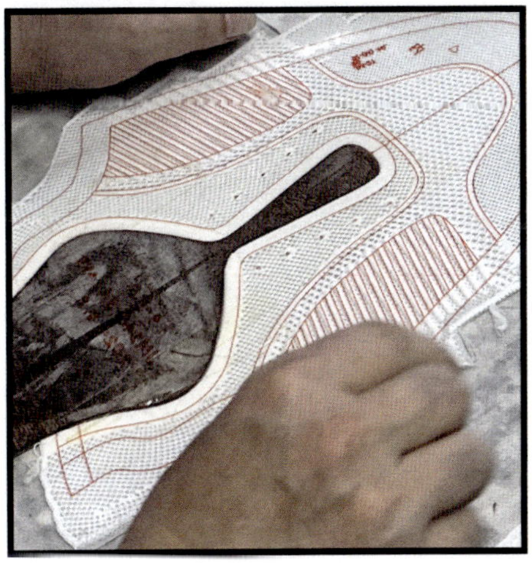

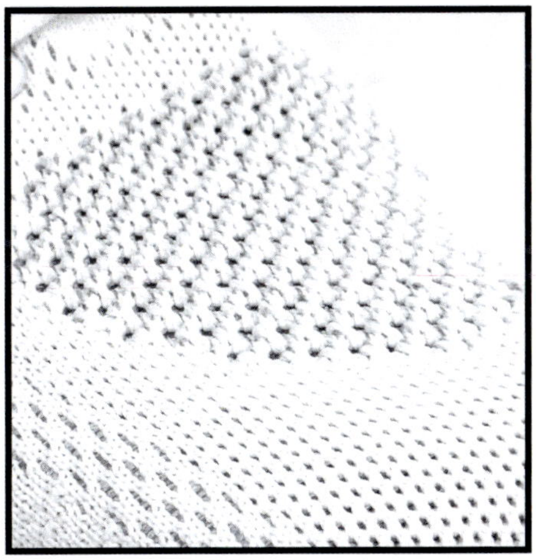

Alpha Edge 4K

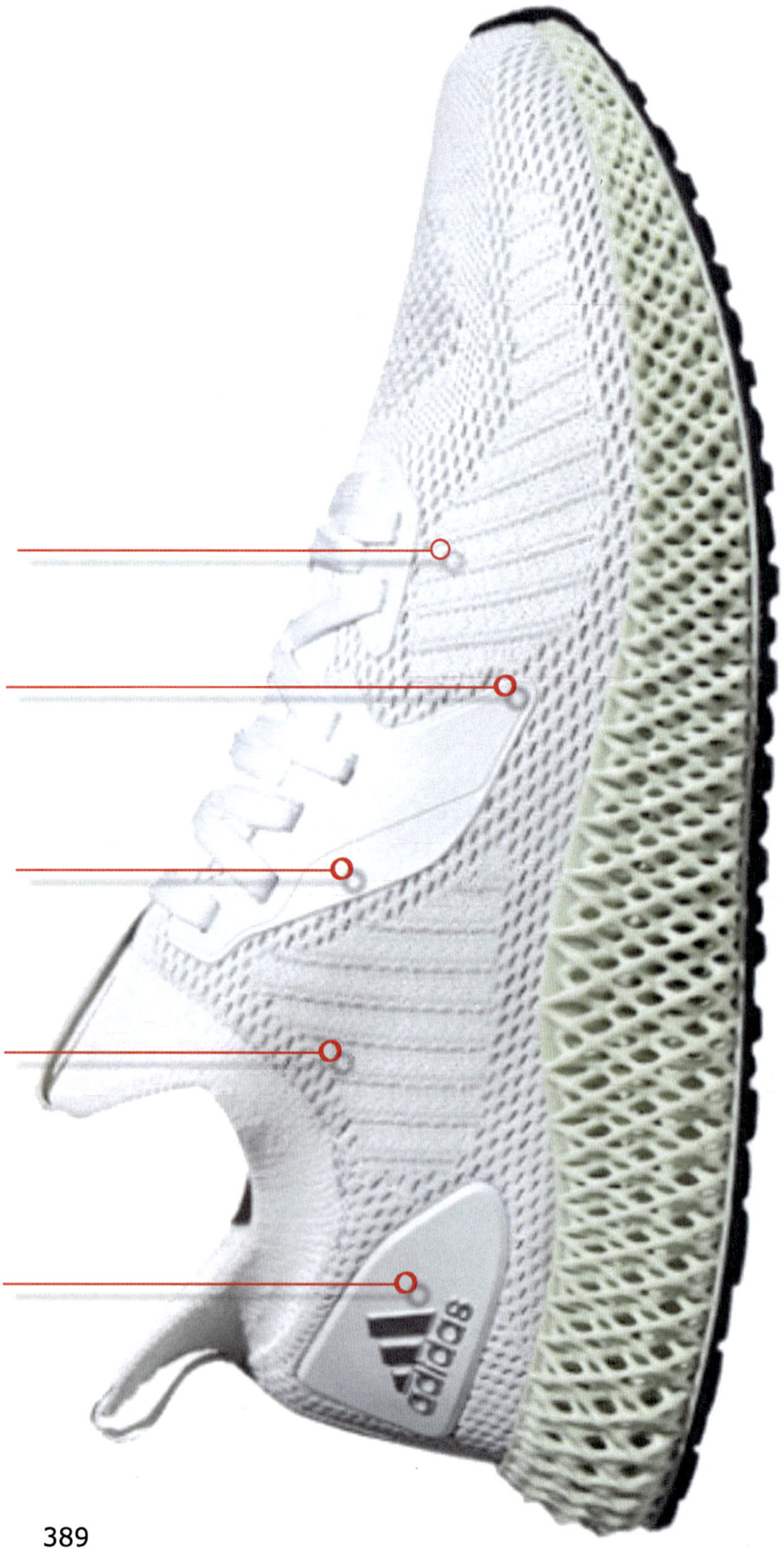

Tooling Cost
16 Sizes = $37,700

Li-Ning ACE + Many items!

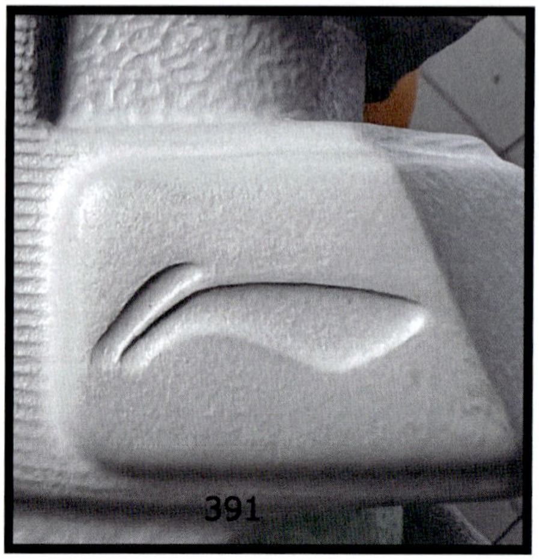

No.
MAKE YOUR OWN WAY

Li-Ning Ace+

Style: Fashion Casual Basketball
Brand: Li-Ning
Model: Wade Essence Ace +
Introduction: 2018
Construction: Strobel
Assembly: Cold cement
Sole unit: Multi-part EVA injection

The Li-Ning Wade Essence is a 'love it or hate it' design. While the Tokyo street styling may not be your thing, everyone can appreciate the combination of design elements and construction techniques. The designers at Li-Ning have broken every rule for efficient footwear production. With 17 different materials in the upper, the Essence looks more like a collage of overstock material rather than a conscience design effort.

The Essence design is made of overlays, windowed underlays, perforations, textures, and color breaks. The designers have specified single needle, double needle, zig-zag, french, rolled, turned out, and butt seams. The shoe has 17 different colors and 8 textures.

The Essence uses almost every design element.

While this shoes looks complicated, it's construction is conventional Strobel with multiple parts of EVA. The sole unit is two parts EVA with molded rubber pods inserted.

Upper:
Cutting dies to make the upper parts
Die cut tools for the footbed
Printing silkscreen for the heel & footbed logos
Molds for rubber logo
Injection mold for strap rings
PU mold for heel bumper

Outsole:
Pad print midsole logos
Rubber press tooling
EVA press tooling

Tongue Lining Logo Silkscreen
1 size x $50 = $50

Back Tab Silkscreen
3 sizes x $50 = $150

Plastic Lasts x 20 sizes
$15 x 350 = $5,250

Upper Cutting Dies
Bent Steel
20 sizes x $150 = $3,000

TPR Strap Logo Molds
3 sizes x $300 = $900

Nylon Strap Ring
Injection Molds
3 sizes x $2,500 = $7,500

Metal Die Casting Tool
1 size x $2400 = $2,400

RF Welding Tool
CNC Cut Brass
+ Silkscreen
1 size x $300 = $300

Polyurethane Mold
Cold Cast resin
5 sizes x $500 = $2,500

Rubber Compression Mold
Cast Steel
20 sizes x $1200= $24,000

EVA Compression Mold
Cast Aluminum
20 sizes x $1500= $30,000

EVA Compression Mold
Cast Aluminum
20 sizes x $1500= $30,000

Footbed Silkscreen
3 sizes x $50 = $150

Outsole Pressing Pads
Cold Cast Urethane
20 sizes x $75 = $1,500

Cold Press EVA Footbed
CNC Cut Steel
6 x $450 = $2,700

Tooling Cost
20 Sizes = $110,850

Pad Print Urethane Fixture
Mini Screen
3 sizes x $50 = $150

Pad print Urethane Fixture
Mini Screen
3 sizes x $50 = $150

Pad Print Urethane Fixture
Mini Screen
3 sizes x $50 = $150

Nike Daybreak

Style: Trainer
Brand: Nike
Model: Daybreak
Introduction: 1975
Construction: Board lasted
Assembly: Cold cement
Sole unit: Cut & Buff EVA

The Nike Daybreak is a classic that helped define the modern athletic shoe. Like the Vans Old Skool, the Nike Daybreak is a time capsule of 70's footwear design and manufacturing technology.

The Daybreak design is the progenitor of all modern running shoes. The design is simply composed of the basic functional elements of a running shoe: logo, heel, toe tip, eyestay, and collar.

Cut and Stitch upper
Requires cutting dies
Heel logo screen

Outsole
Die cut EVA
Rubber pressing mold
Footbed cold press mold
PU pressing pad
Sole grinding stone

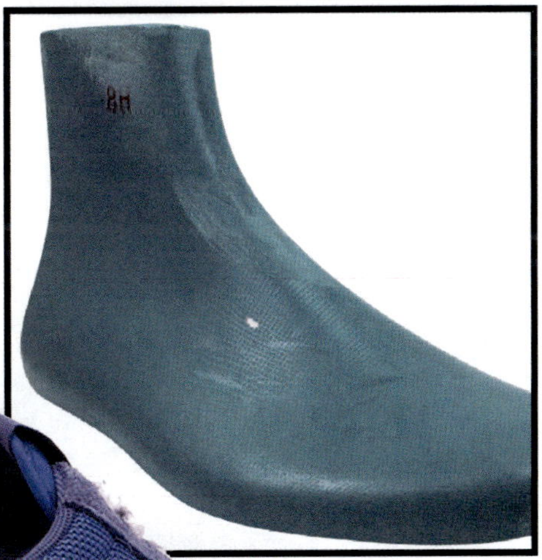

Nike Daybreak

Rubber Compression Mold
Cast Steel
17 sizes x $1000 = $17,000

Plastic Lasts x 17 sizes
$15 x 250 = $3,750

Upper Cutting Dies x 17 sizes
Bent Steel
17 x $150 = $2550

EVA Compression mold 9 sizes
Cast Steel Aluminum
9 sizes x $1500 = $13,500

EVA Midsole Grinding Wheel
4 x $50 = $200

Outsole Pressing Pads
Cold Cast Urethane
17 sizes x $75 = $1,275

Logo Screen Print
5 sizes x $50 = $250

In the next chapter we will fill in this tooling budget.

Nike Daybreak Tooling Cost

17 Sizes = $38,525

Z18 fx

Nike Day Break

Phase 1	Shoe Development Project Pull over																				
Shoe Size		6	6.5	7	7.5	8	8.5	9	9.5	10	10.5	11	11.5	12	12.5	13	14	15			
Sample room overhead fee								1											1	$500.00	$500.00
Model last development	NEW LAST							4											4	$500.00	$500.00
Lasts needed for testing								1											1	$35.00	$35.00
Upper pattern creation								1											1	$400.00	$400.00
2D CAD outsole drawing								1											1	$175.00	$175.00
3D CAD outsole drawing								1											1	$400.00	$400.00
PU casting tempororary outsole	Resin							4											4	$400.00	$1,600.00
Finished fit test samples								6											6	$72.00	$432.00
Wire transfer fee								2											2	$45.00	$90.00
																					$4,132.00

Phase 2	Shoe Development Project First Size Mold Making																				
Sample mold costs		6	6.5	7	7.5	8	8.5	9	9.5	10	10.5	11	11.5	12	12.5	13	14	15			
Outsole rubber pressing mold	Cast steel flat							1											1	$1,000.00	$1,000.00
EVA Compression midsole	Cast aluminium							4											1	$1,500.00	$1,500.00
EVA Footbed	Compression EVA							1											1	$475.00	$475.00
Outsole pressing fixture	Poured urethane							1											1	$65.00	$65.00
Logo silk screen								1											1	$50.00	$50.00
EVA grinding wheel								1											1	$50.00	$50.00
Upper cutting dies	Steel							1											1	$150.00	$150.00
Working prototypes								6											6	$72.00	$432.00
Wire transfer fee								2											0	$45.00	$0.00
																					$3,722.00

Phase 3	Shoe Development Project Production Equipment Mold Making																				
Size Run US Mens	Type	6	6.5	7	7.5	8	8.5	9	9.5	10	10.5	11	11.5	12	12.5	13	14	15			
Outsole rubber pressing mold	Cast steel	1	1	1	1	1	1	1	1	1	1	1	1	1	1	1	1	1	17	$1,000.00	$17,000.00
EVA Compression midsole	Cast aluminium	1	0	1	0	1	0	0	0	1	0	1	0	1	0	1	1	1	9	$1,500.00	$13,500.00
EVA Footbed	Compression EVA	0	0	0	0	0	0	0	0	0	0	0	0	0	0	0	0	0	0	$475.00	$0.00
Logo silk screen		1						1				1		1		1			5	$50.00	$250.00
EVA grinding wheel		1						1				1				1			4	$50.00	$200.00
Outsole pressing fixture	Poured urethane	1	1	1	1	1	1	1	1	1	1	1	1	1	1	1	1	1	17	$75.00	$1,275.00
Production lasts	Plastic	12	12	12	12	12	12	12	12	24	24	24	12	24	12	12	12	10	250	$15.00	$3,750.00
Credit for 2D and 3D Cad work																			0		$0.00
Production upper cutting dies	Metal	1	1	1	1	1	1	1	1	1	1	1	1	1	1	1	1	1	17	$150.00	$2,550.00
Wire transfer fee																			0	$45.00	$0.00
																					$38,525.00

Phase 4	Shoe Production Cost																				
Size Run US Mens	Type	6	6.5	7	7.5	8	8.5	9	9.5	10	10.5	11	11.5	12	12	13	14	15	Pairs	Cost	Total
		50	50	100	75	50	50	100	75	150	100	200	150	200	200	200	200	200	1500	$ 25.00	$ 37,500.00

Tooling Cost Nike DayBreak +

Ready Accessibility: Investigate

Chapter 31

Building Equipment Budgets

$6,157.00

Phase 3	Shoe Development Project Production Equipment Mold Making																				
Size Run US Mens	Type	6	6.5	7	7.5	8	8.5	9	9.5	10	10.5	11	11.5	12	12.5	13	14	15			
Outsole rubber pressing mold	Cast steel flat	1	1	1	1	1	1	1	1	1	1	1	1	1	1	1	1	1	17	$1,500.00	$25,500.00
PU midsole mold	Cast resin	1	1	1	1	1	1	1	1	1	1	1	1	1	1	1	1	1	17	$500.00	$8,500.00
PU midsole painting fixtire	Cast resin	1	1	1	1	1	1	1	1	1	1	1	1	1	1	1	1	1	17	$50.00	$850.00
Air Bag Mold	CNC cut steel	1				1		1		1		1		1		1		1	8	$500.00	$4,000.00
Air Bag Painting and Cutting fixture	Cast resin	1				1		1		1		1		1		1		1	8	$100.00	$800.00
Injection Side logo "Air Max"	CNC cut steel	1				1		1		1		1				1			6	$2,500.00	$15,000.00
TPR eyestay mold #1	CNC cut aluminium	1				1		1		1		1		1		1		1	8	$300.00	$2,400.00
TPR eyestay mold #2	CNC cut aluminium	1				1		1		1		1		1		1		1	8	$300.00	$2,400.00
TPR Heel "Nike" Mold	CNC cut aluminium	1				1		1		1		1		1		1		1	8	$300.00	$2,400.00
EVA Footbed	Compression EVA	1						1				1				1		1	5	$450.00	$2,250.00
Outsole pressing fixture	Poured urethane	1	1	1	1	1	1	1	1	1	1	1	1	1	1	1	1	1	17	$75.00	$1,275.00
Production lasts	Plastic	12	12	12	12	12	12	12	12	24	24	24	12	24	12	12	12	10	250	$15.00	$3,750.00
Production upper cutting dies	Metal	1	1	1	1	1	1	1	1	1	1	1	1	1	1	1	1	1	17	$150.00	$2,550.00
Credit for 2D and 3D Cad work																			0		$0.00
Wire transfer fee																			0	$45.00	$0.00

$71,675.00

What is your size run?
The size run is the very first thing you need to know when building a tooling budget. This will tell you how many tools you will require*.

*One critical point to understand is that not every design element that requires tooling requires it's own size mold.

What parts cannot share equipment?
The cutting dies required to cut the upper cannot share. Each size will require specific cutting dies. The shoe last cannot be shared. Each size will require it's own last.

Can outsoles be shared?
Sometimes. It is possible for a USA 7 and 7.5 to share outsole tooling but they cannot share a last. Flexible rubber or EVA parts can share but they need to be designed to accommodate this option. For example, the tooling would be a size 7 1/4. This will not work on hard soled shoes or equipment with European sizes as the size spread is too large.

Can injection molded parts be shared?
Yes, and due to the high price of injection molds they should be. In our example shoe above, the side logo is made by injection. Due to its size and location the patterns of several sizes can be standardized to share one part. Footbed molds are almost always shared as they can be cut to follow each last size.

Logos are often shared across sizes
Small rubber logos, die cut logos, and RF welded logos are always shared. These small molds often have high productivity so it's easy to share. In the case of our example, there is no need for 17 sizes of the heel and side logos.

Molded Buckles, webbing loops, and plastic hooks are not graded. The size 5 and 14 will use the same parts.

Upper and outsole fixtures
Computer stitching fixtures must follow the upper pattern and they will be graded per every size. The outsole pressing fixtures also must follow the tooling sizes and they will be one to one.

Silkscreens can be shared depending on the design of the parts.

What is last standardization?
Some parts of the shoe last are shared size-by-size. For example, the heel widths may group grade, so your heel counter parts can share a few sizes.

Women's high-heels for example, can share. To make a complete size run of high-heel shoes you will only need three sizes of heel counters.

Nike Air Max

Style: Running
Brand: Nike
Model: Air Max 90
Introduction: 1990
Construction: Strobel lasted
Assembly: Cold cement
Sole unit: Poured PU with airbag and rubber sole.

The Air Max 90 was known as the Air Max III until 2000 when it was reissued taking its name from the year of its launch. The original colorway of white/black/cool grey with infrared was chosen to exaggerate the thickness of the sole air cushion.

The Air Max III was a state of the art running shoe the year of its release. It's now a staple, overshadowing many of the less memorable Air Max models that followed.

Tooling Required to make the Air Max 90:
Upper:
Cutting dies to create the upper
Micro injection tooling for TPR eyestay and heel logos
Standard Injection for "Air Max" side panel logos.

Outsole:
Two-color rubber outsole bottom compression molds
Airbag blow mold clamping tool
PU pouring midsole mold with airbag mounts
Airbag cutting & painting fixtures

TPR Eyestay Mold
8 sizes x $300 = $2,400
Plastic Lasts x 17 sizes
$15 x 250 = $3,750
TPR Eyestay Mold
8 sizes x $300 = $2,400
Injection Mold
CNC Steel
6 x $2500 = $15,000
Upper Cutting Dies
17 Sizes Bent Steel
17 x $150 = $2550
TPR Heel Logo Mold
8 sizes x $300 = $2,400
AIR MAX

Nike Air Max

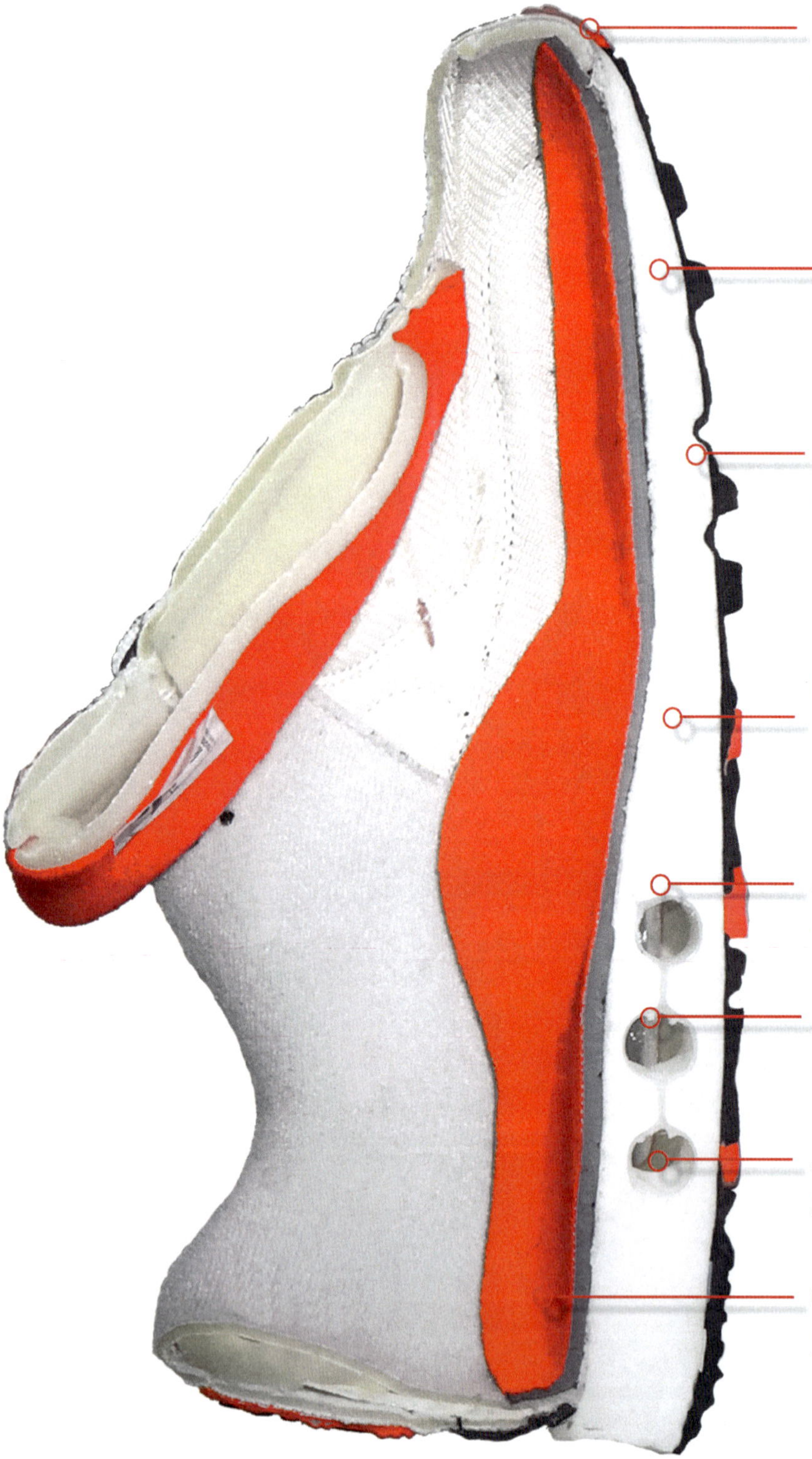
Rubber Compression Mold
Cast Steel
17 sizes x $1500 = $25,500
Polyurethane Midsole x 17 sizes
Cast Resin Molds
17 x $500 = $8,500
Outsole Pressing Pads
Cold Cast Urethane
17 sizes x $75 = $1,275
Midsole Paint Fixture x 17 sizes
17 x $50 = $850
Airbag Compression Mold
8 x $500 = $4,000
Airbag Cutting Fixture
8 x $50 = $400
Airbag Painting Fixture
8 x $50 = $400
Cold Press EVA Footbed x 5 sizes
5 x $450 = $2,250

Nike Air Max Tooling Cost

17 Sizes = $71,675

Nike Air Max

Phase 1	Shoe Development Project Pull over																				
Shoe Size		6	6.5	7	7.5	8	8.5	9	9.5	10	10.5	11	11.5	12	12.5	13	14	15			
Sample room overhead fee								1											1	$500.00	$500.00
Model last development	NEW LAST							4											4	$500.00	$500.00
Lasts needed for testing								1											1	$35.00	$35.00
Upper pattern creation								1											1	$400.00	$400.00
2D CAD outsole drawing								1											1	$175.00	$175.00
3D CAD outsole drawing								1											1	$400.00	$400.00
PU casting temproporary outsole	Resin							4											4	$400.00	$1,600.00
Finished fit test samples								6											6	$72.00	$432.00
Wire transfer fee								2											2	$45.00	$90.00
																					$4,132.00

Phase 2	Shoe Development Project First Size Mold Making																				
Sample mold costs		6	6.5	7	7.5	8	8.5	9	9.5	10	10.5	11	11.5	12	12.5	13	14	15			
Outsole rubber pressing mold	Cast steel flat							1											1	$1,500.00	$1,500.00
PU midsole mold	Cast resin							1											1	$500.00	$500.00
PU midsole painting fixtire	Cast resin							1											1	$50.00	$50.00
Air Bag Mold	CNC cut steel							1											1	$500.00	$500.00
Air Bag Painting and Cutting fixture	Cast resin							1											1	$100.00	$100.00
Injection Side logo "Air Max"	CNC cut steel							1											1	$1,500.00	$1,500.00
TPR eyestay mold #1	CNC cut aluminium							1											1	$300.00	$300.00
TPR eyestay mold #2	CNC cut aluminium							1											1	$300.00	$300.00
TPR Heel "Nike" Mold	CNC cut aluminium							1											1	$300.00	$300.00
EVA Footbed	Compression EVA							1											1	$450.00	$450.00
Outsole pressing fixture	Poured urethane							1											1	$75.00	$75.00
Upper cutting dies	Steel							1											1	$150.00	$150.00
Working prototypes								6											6	$72.00	$432.00
Wire transfer fee								2											0	$45.00	$0.00
																					$6,157.00

Phase 3	Shoe Development Project Production Equipment Mold Making																				
Size Run US Mens	Type	6	6.5	7	7.5	8	8.5	9	9.5	10	10.5	11	11.5	12	12.5	13	14	15			
Outsole rubber pressing mold	Cast steel flat	1	1	1	1	1	1	1	1	1	1	1	1	1	1	1	1	1	17	$1,500.00	$25,500.00
PU midsole mold	Cast resin	1	1	1	1	1	1	1	1	1	1	1	1	1	1	1	1	1	17	$500.00	$8,500.00
PU midsole painting fixtire	Cast resin	1	1	1	1	1	1	1	1	1	1	1	1	1	1	1	1	1	17	$50.00	$850.00
Air Bag Mold	CNC cut steel	1				1		1		1		1		1		1		1	8	$500.00	$4,000.00
Air Bag Painting and Cutting fixture	Cast resin	1				1		1		1		1		1		1		1	8	$100.00	$800.00
Injection Side logo "Air Max"	CNC cut steel	1				1		1		1		1				1			6	$2,500.00	$15,000.00
TPR eyestay mold #1	CNC cut aluminium	1				1		1		1		1		1		1		1	8	$300.00	$2,400.00
TPR eyestay mold #2	CNC cut aluminium	1				1		1		1		1		1		1		1	8	$300.00	$2,400.00
TPR Heel "Nike" Mold	CNC cut aluminium	1				1		1		1		1		1		1		1	8	$300.00	$2,400.00
EVA Footbed	Compression EVA	1						1				1				1		1	5	$450.00	$2,250.00
Outsole pressing fixture	Poured urethane	1	1	1	1	1	1	1	1	1	1	1	1	1	1	1	1	1	17	$75.00	$1,275.00
Production lasts	Plastic	12	12	12	12	12	12	12	12	24	24	24	12	24	12	12	12	10	250	$15.00	$3,750.00
Production upper cutting dies	Metal	1	1	1	1	1	1	1	1	1	1	1	1	1	1	1	1	1	17	$150.00	$2,550.00
Credit for 2D and 3D Cad work																			0		$0.00
Wire transfer fee																			0	$45.00	$0.00
																					$71,675.00

Nike Air Jordan 1

Style: Basketball / Casual
Brand: Nike / Jordan
Model: Air Jordan 1
Introduction: 1982
Construction: Strobel lasted
Assembly: Cold cement
Sole unit: Poured PU air cell with compression rubber cupsole.

The Air Jordan 1 is the modern classic basketball sneaker. The Air Jordan 1 is a simple cut and stitch construction. The original release was made of die cut leather.

Tooling required to make the Jordan 1:
Upper:
Cutting dies to create the upper
RF welding logo tool

Outsole:
Two-color rubber outsole bottom compression molds
Airbag blow mold clamping tool
PU pouring midsole mold with airbag mounts

Plastic Lasts
x 18 sizes
$15 x 350 = $5,250
Upper Cutting Dies
x 18 sizes
Bent Steel
18 x $150 = $2,700
RF Welding Tool
CNC Cut Brass
+ Silkscreen
6 sizes x $300 = $1,800
AIR JORDAN

Nike Air Jordan 1

Nike Air Jordan 1 Tooling Cost

18 Sizes = $54,750

Nike Air Jordan 1

Phase 1	Shoe Development Project Pull over																					
Shoe Size		7	7.5	8	8.5	9	9.5	10	11	11	11.5	12	12.5	13	14	15	16	17	18			
Sample room overhead fee								1												1	$500.00	$500.00
Model last development	NEW LAST							4												4	$500.00	$500.00
Lasts needed for testing								1												1	$35.00	$35.00
Upper pattern creation								1												1	$400.00	$400.00
2D CAD outsole drawing								1												1	$175.00	$175.00
3D CAD outsole drawing								1												1	$400.00	$400.00
PU casting tempororary outsole	Resin							4												4	$400.00	$1,600.00
Finished fit test samples								6												6	$72.00	$432.00
Wire transfer fee								2												2	$45.00	$90.00
																						$4,132.00

Phase 2	Shoe Development Project First Size Mold Making																					
Sample mold costs		7	7.5	8	8.5	9	9.5	10	11	11	11.5	12	12.5	13	14	15	16	17	18			
Outsole rubber pressing mold	Cast steel flat							1												1	$1,500.00	$1,500.00
PU midsole mold	Cast resin							1												1	$500.00	$500.00
Air Bag Mold	CNC cut steel							1												1	$500.00	$500.00
Air Bag Painting and Cutting fixture	Cast resin							1												1	$100.00	$100.00
Injection Side logo "Air Max"	CNC cut steel							1												1	$1,500.00	$1,500.00
TPR eyestay mold #1	CNC cut aluminium							1												1	$300.00	$300.00
TPR eyestay mold #2	CNC cut aluminium							1												1	$300.00	$300.00
TPR Heel "Nike" Mold	CNC cut aluminium							1												1	$300.00	$300.00
EVA Footbed	Compression EVA							1												1	$450.00	$450.00
Outsole pressing fixture	Poured urethane							1												1	$75.00	$75.00
Upper cutting dies	Steel							1												1	$150.00	$150.00
Working prototypes								6												6	$72.00	$432.00
Wire transfer fee								2												0	$45.00	$0.00
																						$6,107.00

Phase 3	Shoe Development Project Production Equipment Mold Making																					
Size Run US Mens	Type	7	7.5	8	8.5	9	9.5	10	11	11	11.5	12	12.5	13	14	15	16	17	18			
Outsole rubber pressing mold	Cast steel flat	1	1	1	1	1	1	1	1	1	1	1	1	1	1	1	1	1	1	18	$1,500.00	$27,000.00
PU midsole mold	Cast resin	1	1	1	1	1	1	1	1	1	1	1	1	1	1	1	1	1	1	18	$500.00	$9,000.00
Air Bag Mold	CNC cut steel	1				1		1		1		1		1		1		1	1	9	$500.00	$4,500.00
Air Bag Cutting fixture	Bent Metal	1				1		1		1		1		1		1		1	1	9	$50.00	$450.00
RF Welding Tool	CNC cut brass	1				1		1		1		1				1				6	$300.00	$1,800.00
EVA Footbed	Compression EVA	1						1				1				1		1	1	6	$450.00	$2,700.00
Outsole pressing fixture	Poured urethane	1	1	1	1	1	1	1	1	1	1	1	1	1	1	1	1	1	1	18	$75.00	$1,350.00
Production lasts	Plastic	12	12	12	12	24	24	24	24	24	24	24	24	24	24	24	14	12	12	350	$15.00	$5,250.00
Production upper cutting dies	Metal	1	1	1	1	1	1	1	1	1	1	1	1	1	1	1	1	1	1	18	$150.00	$2,700.00
Credit for 2D and 3D Cad work																				0		$0.00
Wire transfer fee																				0	$45.00	$0.00
																						$54,750.00

Phase 4	Shoe Production Cost																					
Size Run US Mens	Type	7	7.5	8	8.5	9	9.5	10	11	11	11.5	12	12.5	13	14	15	16	17	18	Pairs	Cost	Total
		50	50	100	75	50	50	100	75	150	100	200	150	200	200	200	200	200	200	1500	$ 25.00	$ 37,500.00

Get Out There And Build Your Branded Equipment.

SHOEMAKERS ACADEMY

If you need any help with tooling,
please don't hesitate to contact
"The Shoe Dog " Wade Motawi

TheShoeDog@shoemakersacademy.com